I0034784

Faouzi Derbel, Nabil Derbel, Olfa Kanoun (Eds.)
Power Systems & Smart Energies

Advances in Systems, Signals and Devices

———

Edited by
Olfa Kanoun, University of Chemnitz, Germany

Volume 3

Power Systems & Smart Energies

—

Edited by
Faouzi Derbel, Nabil Derbel, Olfa Kanoun

DE GRUYTER
OLDENBOURG

Editors of this Volume

Prof. Dr.-Ing. Faouzi Derbel
Leipzig University of Applied Sciences
Chair of Smart Diagnostic and Online Monitoring
Wächterstrasse 13
04107 Leipzig, Germany
faouzi.derbel@htwk-leipzig.de

Prof. Dr.-Ing. Olfa Kanoun
Technische Universität Chemnitz
Chair of Measurement and Sensor Technology
Reichenhainer Strasse 70
09126 Chemnitz
olfa.kanoun@etit.tu-chemnitz.de

Prof. Dr.-Eng. Nabil Derbel
University of Sfax
Sfax National Engineering School
Control & Energy Management Laboratory
1173 BP, 3038 SFAX, Tunisia
n.derbel@enis.rnu.tn

ISBN 978-3-11-044615-9
e-ISBN (PDF) 978-3-11-044841-2
e-ISBN (EPUB) 978-3-11-044628-9
Set-ISBN 978-3-11-044842-9
ISSN 2365-7493
e-ISSN 2365-7507

Library of Congress Cataloging-in-Publication Data
A CIP catalog record for this book has been applied for at the Library of Congress.

Bibliographic information published by the Deutsche Nationalbibliothek
The Deutsche Nationalbibliothek lists this publication in the Deutsche Nationalbibliografie;
detailed bibliographic data are available on the Internet at http://dnb.dnb.de.

© 2017 Walter de Gruyter GmbH, Berlin/Boston
Typesetting: Konvertus, Haarlem
Printing and binding: CPI books GmbH, Leck
♾ Printed on acid-free paper
Printed in Germany

www.degruyter.com

Preface of the Volume Editor

The third volume of the Series "Advances in Systems, Signals and Devices" (**ASSD**), contains international scientific articles devoted to the field of power systems and smart energies. The scope of the volume encompasses all aspects of research, development and applications of the science and technology in these fields. The topics concern energy systems and energy transmission, renewable energy systems, hybrid renewable energy systems, photovoltaic systems, solar energy, wind energy, energy storage, batteries, thermal energy, combined heat and thermal power generation, electric machine design, electric machines modelling and control, electrical vehicles, technologies for electro mobility, special machines, variable speed drives, variable speed generating systems, automotive electrical systems, monitoring and diagnostics, electromagnetic compatibility, power systems, transformers, power electronics, topologies and control of power electronic converters.

These fields are addressed by a separate volume of the series. All volumes are edited by a special editorial board made up by renowned scientist from all over the world.

Authors are encouraged to submit novel contributions which include results of research or experimental work discussing new developments in the field of power systems and smart energies. The series can be also addressed for editing special issues for novel developments in specific fields. Guest editors are encouraged to make proposals to the editor in chief of the corresponding main field.

The aim of this international series is to promote the international scientific progress in the fields of systems, signals and devices. It provides at the same time an opportunity to be informed about interesting results that were reported during the international SSD conferences.

It is a big pleasure of ours to work together with the international editorial board consisting of renowned scientists in the field of power systems and smart energies.

The Editors
Faouzi Derbel, Nabil Derbel and Olfa Kanoun

Advances in Systems, Signals and Devices

Series Editor:

Prof. Dr.-Ing. Olfa Kanoun
Technische Universität Chemnitz, Germany.
olfa.kanoun@etit.tu-chemnitz.de

Editors in Chief:

Systems, Automation & Control

Prof. Dr.-Eng. Nabil Derbel
ENIS, University of Sfax, Tunisia
n.derbel@enis.rnu.tn

Power Systems & Smart Energies

Prof. Dr.-Ing. Faouzi Derbel
Leipzig Univ. of Applied Sciences, Germany
faouzi.derbel@htwk-leipzig.de

Communication, Signal Processing & Information Technology

Prof. Dr.-Ing. Faouzi Derbel
Leipzig Univ. of Applied Sciences, Germany
faouzi.derbel@htwk-leipzig.de

Sensors, Circuits & Instrumentation Systems

Prof. Dr.-Ing. Olfa Kanoun
Technische Universität Chemnitz, Germany
olfa.kanoun@etit.tu-chemnitz.de

Editorial Board Members:

Systems, Automation & Control

Dumitru Baleanu, Çankaya University, Ankara, Turkey
Ridha Ben Abdennour, Engineering School of Gabès, Tunisia
Naceur Benhadj, Braïek, ESSTT, Tunis, Tunisia
Mohamed Benrejeb, Engineering School of Tunis, Tunisia
Riccardo Caponetto, Universita' degli Studi di Catania, Italy
Yang Quan Chen, Utah State University, Logan, USA
Mohamed Chtourou, Engineering School of Sfax, Tunisia
Boutaïeb Dahhou, Univ. Paul Sabatier Toulouse, France
Gérard Favier, Université de Nice, France
Florin G. Filip, Romanian Academy Bucharest Romania
Dorin Isoc, Tech. Univ. of Cluj Napoca, Romania
Pierre Melchior, Université de Bordeaux, France
Faïçal Mnif, Sultan qabous Univ. Muscat, Oman
Ahmet B. Özgüler, Bilkent University, Bilkent, Turkey
Manabu Sano, Hiroshima City Univ. Hiroshima, Japan
Abdul-Wahid Saif, King Fahd University, Saudi Arabia
José A. Tenreiro Machado, Engineering Institute of Porto, Portugal
Alexander Pozniak, Instituto Politecniko, National Mexico
Herbert Werner, Univ. of Technology, Hamburg, German
Ronald R. Yager, Mach. Intelligence Inst. Iona College USA
Blas M. Vinagre, Univ. of Extremadura, Badajos, Spain
Lotfi Zadeh, Univ. of California, Berkeley, CA, USA

Power Systems & Smart Energies

Sylvain Allano, Ecole Normale Sup. de Cachan, France
Ibrahim Badran, Philadelphia Univ., Amman, Jordan
Ronnie Belmans, University of Leuven, Belgium
Frdéric Bouillault, University of Paris XI, France
Pascal Brochet, Ecole Centrale de Lille, France
Mohamed Elleuch, Tunis Engineering School, Tunisia
Mohamed B. A. Kamoun, Sfax Engineering School, Tunisia
Mohamed R. Mékidèche, University of Jijel, Algeria
Bernard Multon, Ecole Normale Sup. Cachan, France
Francesco Parasiliti, University of L'Aquila, Italy
Manuel Pérez,Donsión, University of Vigo, Spain
Michel Poloujadoff, University of Paris VI, France
Francesco Profumo, Politecnico di Torino, Italy
Alfred Rufer, Ecole Polytech. Lausanne, Switzerland
Junji Tamura, Kitami Institute of Technology, Japan

Communication, Signal Processing & Information Technology

Til Aach, Achen University, Germany
Kasim Al-Aubidy, Philadelphia Univ., Amman, Jordan
Adel Alimi, Engineering School of Sfax, Tunisia
Najoua Benamara, Engineering School of Sousse, Tunisia
Ridha Bouallegue, Engineering School of Sousse, Tunisia
Dominique Dallet, ENSEIRB, Bordeaux, France
Mohamed Deriche, King Fahd University, Saudi Arabia
Khalifa Djemal, Université d'Evry, Val d'Essonne, France
Daniela Dragomirescu, LAAS, CNRS, Toulouse, France
Khalil Drira, LAAS, CNRS, Toulouse, France
Noureddine Ellouze, Engineering School of Tunis, Tunisia
Faouzi Ghorbel, ENSI, Tunis, Tunisia
Karl Holger, University of Paderborn, Germany
Berthold Lankl, Univ. Bundeswehr, München, Germany
George Moschytz, ETH Zürich, Switzerland
Radu Popescu-Zeletin, Fraunhofer Inst. Fokus, Berlin, Germany
Basel Solimane, ENST, Bretagne, France
Philippe Vanheeghe, Ecole Centrale de Lille France

Sensors, Circuits & Instrumentation Systems

Ali Boukabache, Univ. Paul, Sabatier, Toulouse, France
Georg Brasseur, Graz University of Technology, Austria
Serge Demidenko, Monash University, Selangor, Malaysia
Gerhard Fischerauer, Universität Bayreuth, Germany
Patrick Garda, Univ. Pierre & Marie Curie, Paris, France
P. M. B. Silva Girão, Inst. Superior Técnico, Lisboa, Portugal
Voicu Groza, University of Ottawa, Ottawa, Canada
Volker Hans, University of Essen, Germany
Aimé Lay Ekuakille, Università degli Studi di Lecce, Italy
Mourad Loulou, Engineering School of Sfax, Tunisia
Mohamed Masmoudi, Engineering School of Sfax, Tunisia
Subha Mukhopadhyay, Massey University Turitea, New Zealand
Fernando Puente León, Technical Univ. of München, Germany
Leonard Reindl, Inst. Mikrosystemtec., Freiburg Germany
Pavel Ripka, Tech. Univ. Praha, Czech Republic
Abdulmotaleb El Saddik, SITE, Univ. Ottawa, Ontario, Canada
Gordon Silverman, Manhattan College Riverdale, NY, USA
Rached Tourki, Faculty of Sciences, Monastir, Tunisia
Bernhard Zagar, Johannes Kepler Univ. of Linz, Austria

Contents

R. Haroun, A. El Aroudi, A. Cid-Pastor and L. Martinez-Salamero

Synthesis of a Power Gyrator Based on Sliding Mode Control of two Cascaded Boost Converters Using a Single Sliding Surface

Abstract: In this paper, a systematic method to synthesize a dc power gyrator, based on cascaded connection of two dc-dc converters is introduced. The dc power gyrator is synthesized by means of a sliding-mode control that imposes a proportionality between the input current of the first stage and the output voltage of the second stage. Only one sliding surface is used to drive both switches of the system. The power gyrator based on cascaded connection of two dc-dc converters can be used for obtaining high conversion ratio. A systematic procedure for designing the system is presented and its stability analysis is performed analytically and validated by numerical simulations using PSIM software. As an example of application, it is shown that a gyrator based on the cascade connection of two boost converters can be a good candidate for the impedance matching between a photovoltaic (PV) generator and a resistive load.

Keywords: Power gyrator, Centralized sliding mode control, Impedance matching.

Mathematics Subject Classification 2010: 65C05, 62M20, 93E11, 62F15, 86A22

1 Introduction

The increase in energy sources demand and the related harmful greenhouse gases are topical problems which have prompted to make use of new clean renewable energy sources [1]. The existence of different nature of renewable energy sources (solar, wind, hydropower, etc...) is prompting to increase the use of distributed generation. New trends in the future distribution and generation system shows that the current centralized system will evolve to a new decentralized one in which the consumers will have a role of active stakeholders of the energy generation. In this new scenario, distributed generation systems based on a dc voltage bus seem to have a slight advantage over those based on an ac distributed bus [2].

The preference of a dc voltage bus instead of an ac distribution one is mainly due to the fact that most of the typical consumer loads are supplied in dc. Moreover, most of RES such as PV generators, fuel cells as well as storage batteries and super capacitors use also dc energy. In the dc-based nanogrid context, the future home

R. Haroun, A. El Aroudi, A. Cid-Pastor and L. Martinez-Salamero: Departament d'Enginyeria Electronica, Electrica i Automatica, Escola Tecnica Superior d'Enginyeria, Universitat Rovira i Virgili, Tarragona, Spain, emails: reham.haroun@urv.cat, abdelali.elaroudi@urv.cat, luis.martinez@urv.cat, angel.cid@urv.cat.

De Gruyter Oldenbourg, ASSD – Advances in Systems, Signals and Devices, Volume 3, 2017, pp. 1–17.
DOI 10.1515/9783110448412-001

electric system is expected to have two dc voltage levels: a high-voltage dc (380 V) powering major home appliances and electric vehicle charging and a low-voltage (48 V) for supplying computer loads, low power consumer electronics, lighting etc. [3–5].

Furthermore, the photovoltaic technologies have rapidly expanded during the last decade and it is foreseen that they will increase significantly the proportion of solar energy use which can be considered as the most promising green energy of the new century due to its abundance [6, 7]. The major problem of energy generation from this important energy source is the optimal functioning of the PV panels. This optimization process is traditionally performed by using a Maximum Power Point Tracker (MPPT) [8].

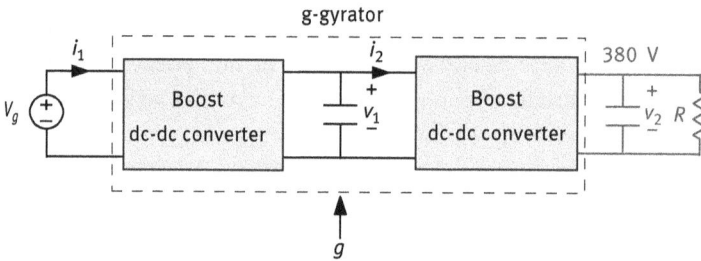

Fig. 1. Power gyrator based on two cascaded boost converters.

In addition, cascaded dc-dc converters (Fig. 1) can be considered as an elemental building block for the future distribution generation systems. As an example, in [6], the authors examined the advantages, difficulties, and implementation issues of using a cascade connection of converters for a series string of PV panels. A typical problem in electrical power generation using PV systems is the high conversion ratio needed to obtain an output dc voltage about 400 V from a low dc voltage of the PV panel. The use of a single stage in performing this conversion ratio will imply working with high duty cycles and therefore will increase the losses which will jeopardize the system efficiency and will reduce the voltage conversion ratio. Therefore, cascaded step-up converters can be a good alternative in terms of efficiency, in order to obtain the desired output voltage [9].

The electrical architecture of power processing systems can be modeled by means of ideal canonical elements and analyzed using the laws governing the interconnection of two-port networks which belong to a class of ideal circuits named POPI (power output is equal to power input) [10]. In [9], the cascaded boost-based converters based gyrator has been synthesized using two sliding surfaces. In addition, the gyrator can be used for impedance matching between a PV panel and a battery as reported in [11]. In this paper, the gyrator based two cascaded boost converters will be used to step up the voltage of the PV panel to the load voltage (380 V). In order to facilitate the design procedure, the gyrator will be synthesized by Sliding-Mode Control (SMC) [12, 13]. This control method has many advantages for controlling these converters such as stability in front of variations of load line, robustness, good dynamic response and simple implementation. It will be shown that a simple strategy based on using only one switching surface for controlling both cascaded converters will be sufficient to achieve our goal.

The rest of the paper is organized as follows: Section II gives an overview about the synthesis of power gyrators using SMC. Section III describes the system under study which consists of a power gyrator based on two cascaded boost converters. The ideal sliding dynamics, the mathematical continuous-time model, its equilibrium point and its stability analysis will be introduced in Section IV. The numerical simulations are performed in Section V. In Section VI, the system is used for impedance matching between a PV panel and a load resistance. Finally, some concluding remarks of this work are summarized in the last section.

2 Synthesis of power gyrators in Sliding-Mode

The concept of power gyrator was introduced in [14, 15] where it was related to a general class of circuits named POPI (power output = power input) describing the ideal behavior of a switched-mode power converter as described in Fig. 2.

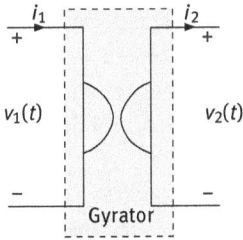

Fig. 2. Schematic of a power gyrator.

A power gyrator is a two-port structure characterized by any of the following two sets of equations

$$I_1 = gV_2, I_2 = gV_1 \qquad \text{g-gyrator} \qquad (1)$$
$$V_1 = rI_2, V_2 = rI_1 \qquad \text{r-gyrator} \qquad (2)$$

where I_1, I_2, V_1 and V_2 are the steady state values of the current and the voltage at the input and output ports respectively and g (r) is the gyrator conductance (resistance).

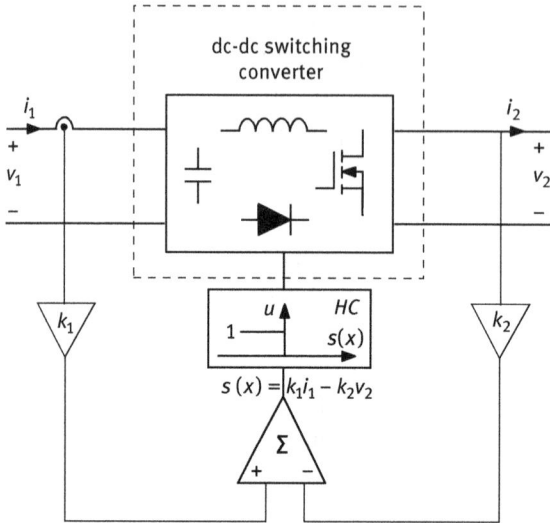

Fig. 3. Block diagram of a power g-gyrator with controlled input current for power processing.

Equations (1) and (2) define the two types of gyrators. The power g-gyrator can be synthesized as shown in the block diagram of Fig. 3. It consists of a switching converter controlled by means of a sliding mode regulation loop [12, 13], in which the switching manifold is the set $\sum = \{x|s(\mathbf{x}) = 0\}$, where $s(\mathbf{x}) = k_1 i_1 + k_2 v_2$ in such a way that, in steady state the following equation holds $I_1 = -(k_2/k_1)V_2 = gV_2$. Note that in this case the gyrator parameter is given by $g = -k_2/k_1$. Imposing a sliding-mode regime requires that the input current i_1 to be a continuous function of time, that implying the existence of a series inductor at the input port. It has to be pointed out that the presence of a Hysteretic Comparator (HC) in the feedback loop of the switching regulator of Fig. 3 will result in a variable switching frequency which will depend mainly on the hysteresis width h and the operating point [16].

The simplest converters with such a constraint are the boost converter and the fourth order structures like boost with output filter (BOF), buck with input filter (BIF), and Ćuk converter with galvanic isolation. It was shown in [17, 18] that both BIF and Ćuk converter, behaving as g-gyrators with controlled output current, can exhibit stable g-gyrator characteristics if capacitive damping are inserted and certain parametric conditions are satisfied. In particular, boost converters, behaving as g-gyrators with controlled input current, can exhibit a sliding regime with unconditionally stable equilibrium point. Therefore, in this study boost converters have been chosen to be cascaded because of their higher efficiency when compared with other fourth order structures having the same sliding and stability characteristics.

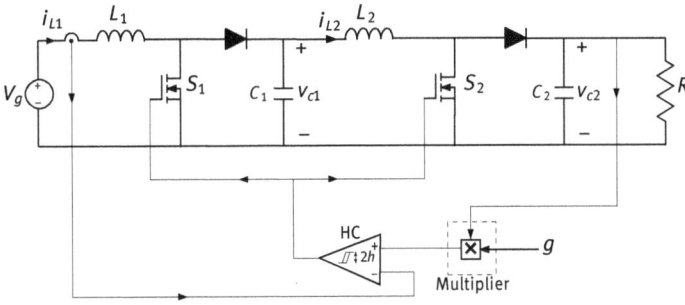

Fig. 4. The schematic diagram for a power gyrator with controlled input current based on two cascaded boost converters.

3 G-Gyrator based on the cascade connection of two boost converters

In this study, our main objective is to design a g-gyrator based on two cascaded boost converters in order to fed a load resistance with a 380 V dc output voltage. For that, the two cascaded converters will be controlled by means of a single sliding surface in order to reduce the number of components of the system. Therefore, a single control variable is used to drive the switches of both subsystems. Figure 4 depicts the circuit description corresponding to g-gyrator based on two cascaded boost converters. The gyrator characteristics can be implemented by means of a sliding mode regulation loop like the described in the previous section. The sliding surface imposes that the input current of the first converter is proportional to the output voltage of the second converter. The sliding surface can be described with the switching function:

$$s(\mathbf{x}) = g v_{c2} - i_{L1}. \tag{3}$$

In steady-state $s(\mathbf{x}) = 0$, i.e., $I_{L1} = g V_{c2}$. Furthermore, by considering that the two converters in Fig. 3 are ideal, one will have $V_g I_{L1} = V_{c2} I_R$ and therefore, (2) will be automatically satisfied to obtain the following relationship between the output current and the input voltage

$$I_R = g V_g \tag{4}$$

4 Ideal sliding dynamics

4.1 Switched model

By applying standard KVLs and KCLs to the circuit depicted in Fig. 4. The two cascaded boost converters can be represented by the following differential equations

$$\frac{di_{L1}}{dt} = \frac{V_g}{L_1} - \frac{(1-u)v_{c1}}{L_1} \tag{5}$$

$$\frac{di_{L2}}{dt} = \frac{v_{c1}}{L_2} - \frac{(1-u)v_{c2}}{L_2} \tag{6}$$

$$\frac{dv_{c1}}{dt} = \frac{(1-u)i_{L1}}{C_1} - \frac{i_{L2}}{C_1} \tag{7}$$

$$\frac{dv_{c2}}{dt} = \frac{(1-u)i_{L2}}{C_2} - \frac{v_{c2}}{RC_2} \tag{8}$$

where V_g is a voltage source which is considered constant. All the other parameters that appear in (5)–(8) are shown in Fig. 4. The signal u is the control variable used to drive the switches of both converters. $u = 1$ during the period T_{ON} and $u = 0$ during the period T_{OFF}.

4.2 Equivalent control

The ideal sliding mode model can be obtained by substituting the discontinuous control variable u, in (5)–(8), which is a binary signal belonging to the set $\{0,1\}$ by its equivalent continuous variable $u_{eq}(\mathbf{x})$ that can take all the values between 0 and 1. This equivalent control variable is obtained by imposing that the trajectories are evolving on the switching manifold. To synthesize a g-gyrator in the two cascaded boost converter stages, the switching function can be selected as follows: $s(\mathbf{x}) = gv_{c2} - i_{L1}$. By imposing the invariance conditions [16], one has $s(\mathbf{x}) = \dot{s}(\mathbf{x}) = 0$, where the overdot stands for the time derivative. Therefore, the dynamical behavior of i_{L1} is constrained by the following differential equation

$$\dot{s}(\mathbf{x}) = g\frac{dv_{c2}}{dt} - \frac{di_{L1}}{dt} = 0 \tag{9}$$

From equations. (5), (8) and (9), the following expression is obtained for the equivalent control variable $u_{eq}(\mathbf{x})$

$$u_{eq} = 1 - \frac{V_gRC_2 + gv_{c2}L_1}{v_{c1}RC_2 + gi_{L2}RL_1} \tag{10}$$

Note that $u_{eq}(\mathbf{x})$ must be bounded by the minimum and maximum value of u [12, 13], i.e

$$0 < u_{eq}(\mathbf{x}) < 1 \tag{11}$$

Substituting (10) in equations (5)–(8) and taking into account (9), the following ideal sliding dynamics reduced-order model is obtained

$$\frac{di_{L2}}{dt} = \frac{v_{c1}}{L_2} - \frac{V_g RC_2 + g v_{c2} L_1}{RL_2(v_{c1} C_2 + g i_{L2} L_1)} v_{c2} \tag{12}$$

$$\frac{dv_{c1}}{dt} = \frac{g(V_g RC_2 + g v_{c2} L_1)v_{c2}}{RC1(v_{c1} C_2 + g i_{L2} L_1)} - \frac{i_{L2}}{C_1} \tag{13}$$

$$\frac{dv_{c2}}{dt} = \frac{V_g RC_2 + g v_{c2} L_1}{RC_2(v_{c1} C_2 + g i_{L2} L_1)} i_{L2} - \frac{v_{c2}}{RC_2} \tag{14}$$

4.3 Equilibrium point

Being a third order nonlinear system, the dynamical analysis of (12)–(14) is challenging. However, the linearization of the system near the operating point reveals that the
system is stable with certain conditions as will be shown later. The equilibrium point
can be obtained by forcing the time derivative of the state variables of the ideal sliding
mode model to be null. From (12)–(14) and taking into account the sliding surface
equation and that the input power equals the output power, the equilibrium point
of the ideal sliding dynamics is given by

$$\mathbf{x}^* = \begin{pmatrix} I_{L1} \\ I_{L2} \\ V_{c1} \\ V_{c2} \end{pmatrix} = \begin{pmatrix} g^2 V_g R \\ g V_g \sqrt{Rg} \\ V_g \sqrt{Rg} \\ g V_g R \end{pmatrix} \tag{15}$$

It can be observed that $I_{L1} = g V_{c2}$ which defines the steady-state gyrator behavior and
that $I_{L2} = g V_{c1}$, which defines a steady-state Loss Free Resistor (LFR) characteristics
that means that the input port of the second boost converter has natural LFR characteristics. The control law at the equilibrium point can be obtained by substituting (15)
in (10). In doing so, one obtains the following steady state value U_{eq} of the control
variable u_{eq}

$$U_{eq} := u_{eq}(\mathbf{x}^*) = 1 - \frac{1}{\sqrt{Rg}} \tag{16}$$

As we mentioned before in (11) that U_{eq} is bounded between 0 and 1, the following
condition should be fulfilled

$$Rg > 1 \tag{17}$$

Tab. 1. Parameter values used in this study.

V_g (V)	V_{c1} (V)	V_{c2} (V)	L_1 (μH)	L_2 (mH)	C_1, C_2 (μF)	g (S)	R (Ω)	h	f_s (kHz)
18	80	380	200	2	10	0.0135	2600	0.1	100

4.4 Stability analysis

In order to study the stability of the system, the nonlinear model (12)–(14) is linearized around the equilibrium point $\mathbf{x}^*$ given by equation (15). The stability of the linearized system can be studied by using the Jacobian matrix $\mathbf{J}$ corresponding to (12)–(14) and evaluating it at the equilibrium point $\mathbf{x}^*$. This matrix can be expressed as follows

$$
\mathbf{J} = \begin{pmatrix}
\dfrac{L_1 g}{L_2 C} & \dfrac{2C_2 + g^2 L_1}{L_2 C} & -\dfrac{2g^2 L_1 + C_2}{\sqrt{RgL_2 C}} \\[2ex]
-\dfrac{2g^2 L_1 + C_2}{C_1 C} & -\dfrac{C_2 g}{C_1 C} & -\dfrac{g(2g^2 L_1 + C_2)}{\sqrt{RgC_1 C}} \\[2ex]
\dfrac{1}{\sqrt{RgC}} & -\dfrac{g}{\sqrt{RgC}} & -\dfrac{1}{RC}
\end{pmatrix}
\tag{18}
$$

where $C = C_2 + g^2 L_1$ and $\Delta\tau^2 = C_2 L_2 - L_1 C_1$. The characteristic polynomial equation of the linearized system can be obtained using the Jacobian matrix $\det(\mathbf{J} - s\mathbf{I}) = 0$, where $\mathbf{I}$ is the unitary matrix. Developing this equation, the characteristic polynomial can be written in the following form

$$
s^3 + \frac{L_2 C_1 + Rg\Delta\tau^2}{RL_2 C_1 C} s^2 + \frac{2L_2 g^2 + 2RgC + C_1}{RgL_2 C_1 C} s + \frac{2}{RL_2 C_1 C}
\tag{19}
$$

The stability of this system can be checked by using Routh-Hurwitz criterion to get the following stability conditions

$$
L_2 C_1 + Rg\Delta\tau^2 > 0
\tag{20}
$$
$$
(L_2 C_1 + Rg\Delta\tau^2)(2g^2 L_2 + C_1) + 2R^2 g^2 C\Delta\tau^2 > 0
\tag{21}
$$

5 Numerical simulations

In order to verify the theoretical results predicted in Section IV, the circuit depicted in Fig. 4 has been simulated by using PSIM software with the set of parameter values depicted in Tab. 1 that satisfies the stability conditions of Section IV. First, the validity of the ideal sliding dynamics model (12)–(14) will be checked using numerical simulation for the full order model. The system is simulated from two certain initial points P_1 and P_2 using the two different models. As shown in Fig. 5, the trajectories of the reduced-order model are in perfect agreement with the full-order model for both cases.

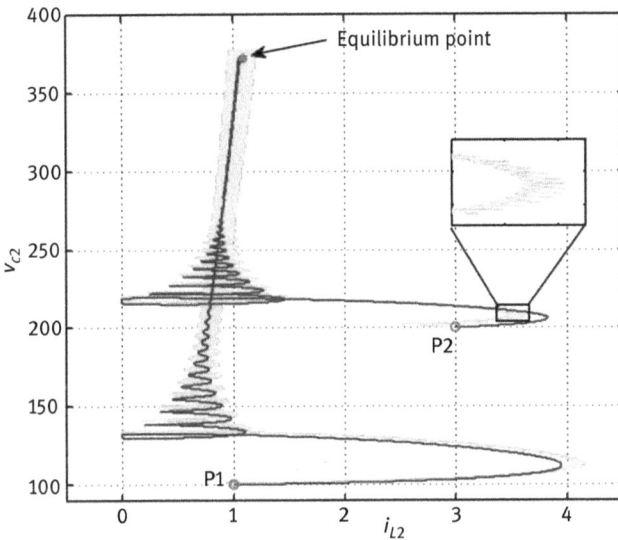

Fig. 5. Trajectories obtained from the reduced-order ideal sliding-mode dynamics model and from the full-order switched model using PSIM starting from different initial conditions P_1 and P_2 in the state plane (i_{L2}, v_{c2}).

Figure 6 shows the transient start up and steady state responses of the system from zero initial conditions. Note, that after a short transient time of 60 ms, the state variables reach their steady state values which are in agreement with (15). The cascade connection of the two converters is behaving as g-gyrator in steady state and the second converter behaves as an LFR in the intermediate point of the two cascaded boost converters, due to the constraints imposed by the switching function $s(\mathbf{x})$.

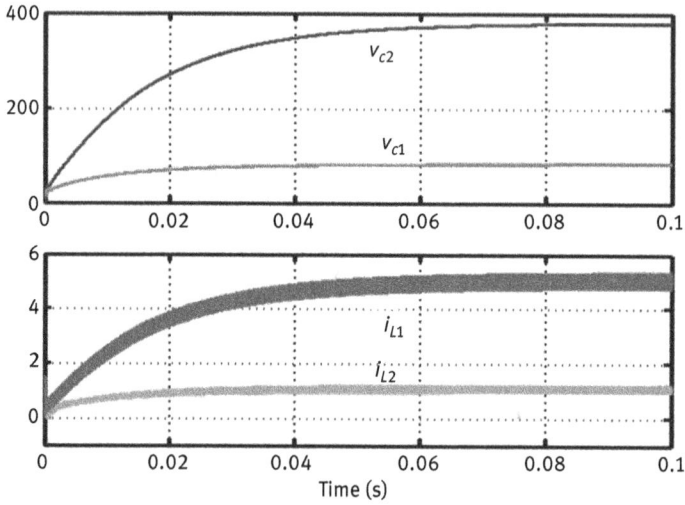

Fig. 6. The capacitor voltages v_{c2}, v_{c1} and the inductor currents i_{L1}, i_{L2} (respectively from up to down) showing the start-up of the g-gyrator with controlled input current based on two cascaded boost converters for $g = 0.013$ S.

Figure 7a shows the effect of the input voltage variation from 18 V to 20 V. By increasing the input voltage, the output capacitor voltages and the inductor currents increase. However, when the output power changes from 90 W (half load) to 180 W (full load)

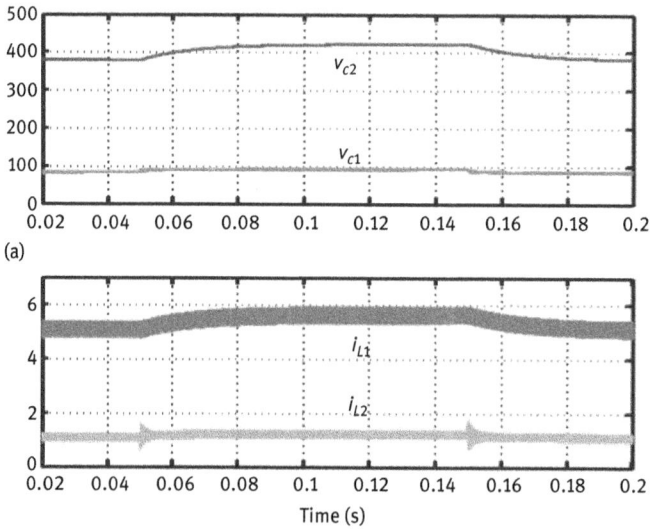

(a)

Fig. 7. The capacitor voltages v_{c2}, v_{c1} and the inductor currents i_{L1}, i_{L2} (respectively from up to down) for g-gyrator with controlled input current based on two cascaded boost converters for $g = 0.013$ S and under (a) Input voltage change from 18 V to 20 V (b) Load change from 90 W to 180 W.

(b)

Time (s)

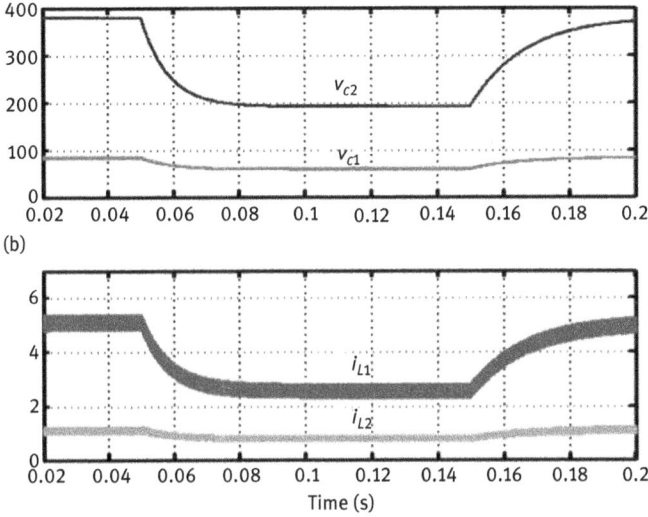

Fig. 7. (continued) The capacitor voltages v_{c2}, v_{c1} and the inductor currents i_{L1}, i_{L2} (respectively from up to down) for g-gyrator with controlled input current based on two cascaded boost converters for $g = 0.013$ S and under (a) Input voltage change from 18 V to 20 V (b) Load change from 90 W to 180 W.

(the load R changes from 1600 to 800 Ω), the capacitor voltages and the inductor currents decrease as shown in Fig. 7b. This can also be deduced from (15).

6 Impedance matching between a PV panel and load using power gyrator based on cascaded boost converters

In this section, the gyrator based on two cascaded boost converters of the previous sections will be supplied by a PV panel as illustrated in Fig. 8. The used PV panel is BP585 and the required voltage is 380 V at the output as in the previous section. The used hysteresis width is $h = 0.1$ A which results in a steady-state switching frequency f_s of about 100 kHz.

The gyrator has been used as interfacing element between the PV panel and the load. This can be considered as an example of application in impedance matching in PV systems when supplying a load resistance with 380 V. The circuit is simulated using PSIM and the parameter g for the control can be obtained from the MPPT controller to supply the two boost converters with the same control variable. An extremum-seeking control MPPT is used to extract the maximum power from the PV panel [8].

The direct connection of the PV panel to the load resulting in the operating point A as shown in Fig. 9. The variation of the parameter g of the gyrator changes the

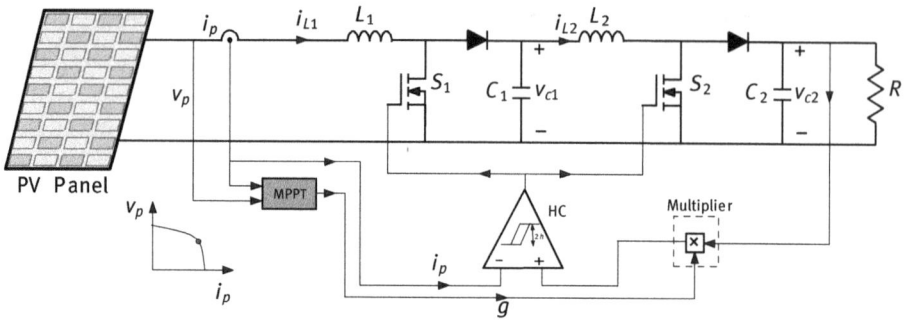

Fig. 8. Impedance matching of a PV panel by means of two cascaded boost converters.

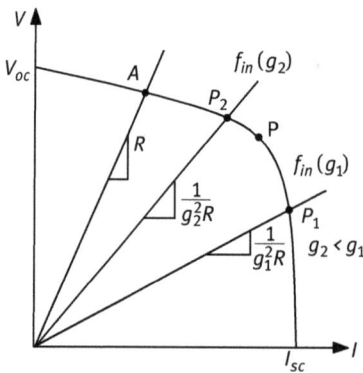

Fig. 9. Impedance matching between a PV panel and a resistive load.

operating point of the PV panel as illustrated in Fig. 9. Operating points P_1 and P_2 are corresponding respectively to conductances g_1 and g_2 with $g_2 < g_1$. The goal of the matching is to find an optimal value of the conductance g that leads to an intersection of the PV V-I characteristic and $f_{in}(g)$ characteristic at the maximum power point which can be defined with the operating point P in Fig. 9.

Figure 10 shows the steady state behavior for the gyrator based on two cascaded boost converters supplied from a PV panel using an MPPT controller. It can be noticed that the output voltage and the output current for the PV panel are $180°$ out of phase. The frequency of the instantaneous power p is twice the frequency of current or voltage. Therefore, each half period of current or voltage, a maximum value of p is reached. The intermediate voltage v_{c1} has the same frequency as the voltage of the PV panel and its value is around 80 V. Finally, the output voltage v_{c2} has the same frequency of the input current and its averaged value is 380 V.

The response of the gyrator based on two cascaded boost converters connected to the PV panel with an MPPT have been checked also under the change of temperature T and irradiance S. The PV $i-v$ characteristic curve and the gyrator load line are depicted

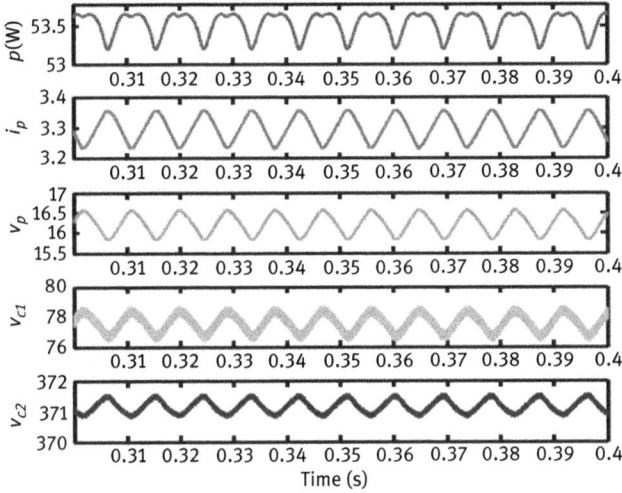

Fig. 10. Steady state waveforms of the power gyrator based on two cascaded boost converters and supplied from a PV panel with an MPPT controller.

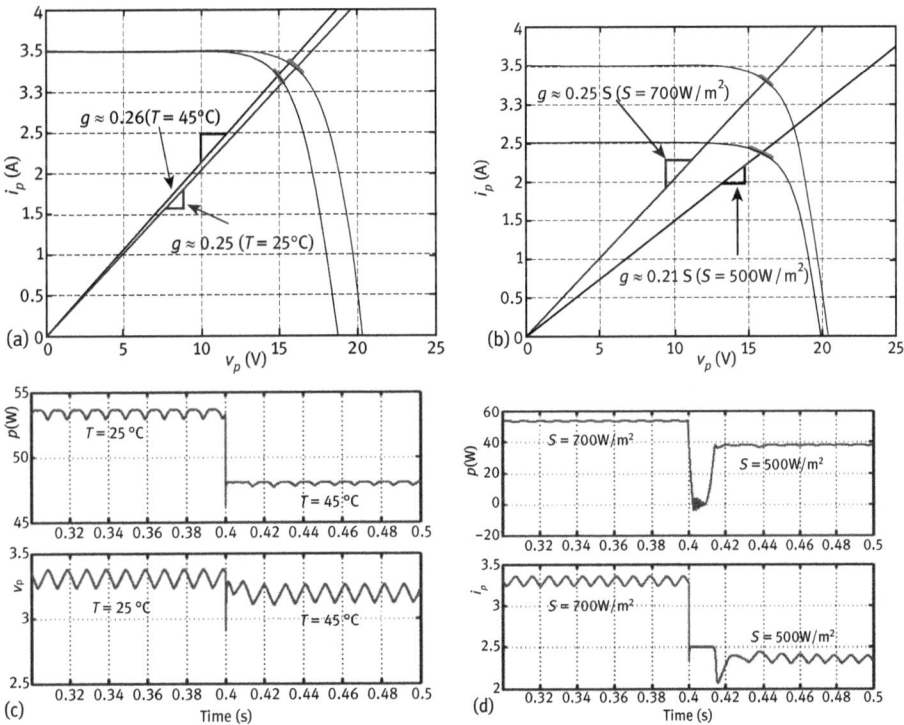

Fig. 11. Response of the the gyrator based on two cascaded boost converters supplied by the PV panel with the MPPT controller under temperature and irradiation changes. (a): Temperature effect at $S = 700$ W/m^2. (b): Irradiance effect at $T = 25^oC$. (c): Temperature effect at $S = 700$ W/m^2. (d): Irradiance effect at $T = 25^oC$.

in Fig. 11a and Fig. 11b for the same step change in the temperature and irradiance respectively. It can be noticed that, when the temperature increases from 25 °C to 45 °C, the average conductance g changes from 0.25 S to 0.26 S to extract the maximum power as shown in Fig. 11a. When the irradiance changes from $S = 700$ W/m^2 to $S = 500$ W/m^2, the conductance g changes from 0.25 S to 0.21 S for achieving the Maximum Power Point (MPP) steady state for the new conditions as shown in Fig. 11b. Therefore, the steady-state of the system for both irradiance levels is oscillating around the MPP. This can also be observed in the corresponding waveforms of v_p and p which are depicted in Fig. 11c, 11d respectively. It can be noticed that when the temperature increases, the power decreases as shown in Fig. 11c. Similarly, when the irradiance decreases, the power decreases as shown in Fig. 11d, while maintaining the operating point at the MPP for the both cases.

7 Conclusion

In this paper, the gyrator concept has been used to connect two cascaded boost converters using a single sliding surface to reduce the number of components which will consequently decrease the cost and increase the efficiency. This system has been analyzed theoretically by means of numerical simulation using PSIM. The ideal sliding model has been obtained and validated using the full-order switched model. Stability analysis has been carried out for the gyrator based on two cascaded boost converters and the conditions for sliding motions and for stability have been derived. Using the gyrator canonical element based on SMC adds simplicity for the stability analysis and the implementation. It has been deduced that the gyrator based two cascaded boost converters has a natural LFR in the intermediate point of the converters. It has also been shown that gyrator based cascaded boost converters can be used to solve the problem of achieving high voltage conversion ratio and at the same time for dc impedance matching in photovoltaic systems with potentially a better efficiency if only a single switching surface is used to drive the switches of the two cascaded converters.

Acknowledgements: This work was supported by the Spanish MINECO under grants DPI2010-16481, DPI2010-16084 and CSD2009-00046.

Bibliography

[1] A. Cellatoglu and K. Balasubramanian. Renewable energy resources for residential applications in coastal areas: A modular approach. 42nd *Southeastern Symp. on System Theory (SSST)*, 2010.

[2] S. Luo and I. Bataresh. A Review of distributed power systems part I: DC distributed power system. *IEEE Aerospace and Electronic Systems Magazine*, 20(8):5–16, 2005.

[3] D. Salomonsson and A. Sannino. Low-voltage DC distribution system for commercial power systems with sensitive electronic loads. *IEEE Trans. on Power Delivery*, 22(3):1620– 1627, 2007.

[4] P. Karlsson and J. Svensson. DC bus voltage control for a distributed power system. *IEEE Trans. on Power Electronics*, 18(6):1405– 1412, 2003.

[5] M. Brenna, G. Lazaroiu and E. Tironi. High power quality and DG integrated low voltage dc distribution system. in *IEEE Power Engineering Society General Meeting*, 2006.

[6] G. R. Walker and P. C. Sernia. Cascaded DC-DC converter connection of photovoltaic modules. *IEEE Trans. on Power Electronics*, 19(7):1130–1139, 2004.

[7] A. I. Bratcu, I. Munteanu, S. Bacha, D. Picault and B. Raison. Cascaded DC-DC converter photovoltaic systems: power optimization issues. *IEEE Trans. on Industrial Electronics*, 58(2):403–411, 2011.

[8] R. Leyva, C. Alonso, I. Queinnec, A. Cid-Pastor, D. Lagrange and L. Martinez-Salamero. MPPT of photovoltaic systems using extremum-seeking control. *IEEE Trans. on Aerospace and Electronic Systems*, 42(1):249–258, 2006.

[9] R. Haroun, A. Cid-Pastor, A. El Aroudi, and L. Martinez-Salamero. Synthesis of canonical elements for power processing in dc distribution systems using cascaded converters and sliding-mode control. *IEEE Trans. on Power Electronics*, Early acsess 2013.

[10] S. Singer and R. Erickson. Canonical modeling of power processing circuits based on the POPI concept. *IEEE Trans. on Power Electronics*, 7(1):37–34, 1992.

[11] A. Cid-Pastor, L. Martinez-Salamero, C. Alonso, G. Schweitz and R. Leyva. DC Power Gyrator versus DC Power Transformer for Impedance Matching of a PV Array. 12th *Int. Power Electronics and Motion Control Conf. (EPE-PEMC)*, :1853– 1858, Aug. 2006.

[12] V. I. Utkin. *Sliding modes and their application in variable structure systems*. MIR Publishers, 1978.

[13] P. Mattavelli, L. Rossetto, G. Spiazzi, and P. Tenti. General-purpose sliding-mode controller for DC/DC converter applications. 24th *Annual IEEE Power Electronics Specialists Conf., (PESC)*, :609–615, 1993.

[14] S. Singer. Gyrators Application in Power Processing Circuits. *IEEE Trans. on Industrial Electronics*, 34(3):313–318, 1987.

[15] S. Singer. Loss-free gyrator realization. *IEEE Trans. on Circuits and Systems*, 35(1):26–34, 1988.

[16] H. Sira-Ramirez. Sliding motions in bilinear switched networks. *IEEE Trans. on Circuits and Systems*, 34(8):919–933, 1987.

[17] A. Cid-Pastor, L. Martinez-Salamero, C. Alonso, G. Schweitz, J. Calvente, and S. Singer. Classification and synthesis of power gyrators. in *IEE Proc. on Electric Power Applications*, 153:802–808, 2006.

[18] A. Cid Pastor, *Energy processing by means of power gyrators*. PhD thesis, Technical University of Catalonia (UPC), Barcelona, July 2005.

Biographies

Reham Haroun was born in Egypt in 1982. She obtained the graduate degree in power and electrical engineering from Aswan Faculty of Engineering, South Valley University, Aswan, Egypt, in 2004 and the Master degree from the same University in 2009 where she worked as lecture assistant during the period 2004–2009. During the same period, she was a member of Aswan Power Electronics Application Research Center (APEARC) group. Her research interests are in power electronics applications including dc-dc switched power supply and AC-DC Power Factor Correction (PFC) converters. She received her Ph.D degree at Universitat Rovira i Virgili, Tarragona, Spain in 2014.

Abdelali El Aroudi was born in Tangier, Morocco, in 1973. He received the graduate degree in physical science from Faculté des sciences, Université Abdelmalek Essaadi, Tetouan, Morocco, in 1995, and the Ph.D. degree (hons) from Universitat Politècnica de Catalunya, Barcelona, Spain in 2000. During the period 1999–2001 he was a Visiting Professor at the Department of Electronics, Electrical Engineering and Automatic Control, Technical School of Universitat Rovira i Virgili (URV), Tarragona, Spain, where he became an associate professor in 2001 and a full-time tenure Associate Professor in 2005. From September 2007 to January 2008 he was holding a visiting scholarship at the Department of Mathematics and Statistics, Universidad Nacional de Colombia, Manizales, conducting research on modeling of power Electronics circuits for energy management. From February 2008 to July 2008, he was a visiting scholar at the Centre de Recherche en Sciences et Technologies de Communications et de l'Informations (CReSTIC), Reims, France. He has participated in three Spanish national research projects and five cooperative international projects. His research interests are in the field of structure and control of power conditioning systems for autonomous systems, power factor correction, stability problems, nonlinear phenomena, chaotic dynamics, bifurcations and control. He serves as usual reviewer for many scientific journals. He has published more than 150 papers in scientific journals and conference proceedings. He is a member of the GAEI research group (Universitat Rovira i Virgili) on Industrial Electronics and Automatic Control whose main research fields are power conditioning for vehicles, satellites and renewable energy. He has given invited talks in several universities in Europe, South America and Africa.

Angel Cid-Pastor graduated as Ingeniero en Electrónica Industrial in 1999 and as Ingeniero en Automàtica y Electrónica Industrial in 2002 at Universitat Rovira i Virgili, Tarragona, Spain. He received the M.S. degree in design of microelectronics and microsystems circuits in 2003 from Institut National des Sciences Appliquées, Toulouse, France. He received the Ph.D. degree from Universitat Politècnica de Catalunya, Barcelona, Spain, and from Institut National des Sciences Appliquées, LAAS-CNRS Toulouse, France in 2005 and 2006, respectively. He is currently an Associated Professor at the Departament d'Enginyeria Electrònica, Elèctrica i Automàtica, Escola Tècnica Superior d'Enginyeria, Universitat Rovira i Virgili, Tarragona, Spain. His research interests are in the field of power electronics and renewable energy systems.

Luis Martinez-Salamero received the Ingeniero de Telecomunicación and the doctorate degrees from the Universidad Politécnica de Cataluña, Barcelona, Spain in 1978 and 1984, respectively. From 1978 to 1992, he taught circuit theory, analog electronics and power processing at Escuela Técnica Superior de Ingenieros de Telecomunicación de Barcelona. During the academic year 1992–1993 he was visiting professor at the Center for Solid State Power Conditioning and Control, Deparment of Electrical Engineering, Duke University, Durham, NC. He is currently a Full Professor at the Departamento de Ingeniería Electrónica, Eléctrica y Automática, Escuela Técnica Superior de Ingeniera, Universidad Rovira i Virgili, Tarragona, Spain. During the academic years 2003–2004 and 2010–2011, he was a Visiting Scholar at the Laboratoire d'Architecture et d'Analyse des Systèmes (LAAS) of the Research National Center (CNRS), Toulouse, France. His research interest are in the field of structure and control of power conditioning systems for autonomous systems. He has published a great number of papers in scientific journals and conference proceedings and holds a US patent on the electric energy distribution in vehicles by means of a bidirectional dc-to -dc switching converter. He is the director of the GAEI, research group on Industrial Electronics and Automatic Control whose main research fields are power conditioning for vehicles, satellites and renewable energy. Dr Martínez-Salamero was Guest Editor of the IEEE Trans. on Circuits And Systems (1997) for the special issue on Simulation, Theory and Design of Switched-Analog Networks. He has been distinguished lecturer of the IEEE Circuits and Systems Society in the period 2001–2002.

F. Flores-Bahamonde, H. Valderrama-Blavi, J. A. Barrado Rodrigo,
J. M. Bosque and A. Leon-Masich

Evaluating Power Converters using a Wind-System Simulator

Abstract: Distributed systems become an important solution to the generation of green energy integrating different renewable energy sources. The randomness of this kind of sources is counteracted by storage elements, injecting the energy into the grid according to a given power profile. Integration of wind generator to a DC-bus micro-grid requires the study of three-phase rectifiers to inject the energy, delivered by the generator, into the DC-bus. Different converter topologies can be proposed for both power factor capability, and maximum power point tracking (MPPT) algorithm. To optimize the overall system, a wind-system simulator is required to make repeatable experiments for testing the performance of three different power rectifiers. In addition, MPPT algorithms can also be examined. This paper considers the development of such experimental tool using the graphical environment Labview. This simulator is programmed to drive,by means of a torque open-loop, an induction machine mechanically coupled to the wind generator. Finally, the evaluations of the three-phase PFC rectifiers are carried out, and a classical MPPT algorithm is also implemented to test the simulator.

Keywords: Wind energy, Wind-system simulator, PFC rectifiers, Three-phase LFR, MPPT algorithm.

Mathematics Subject Classification 2010: 65C05, 62M20, 93E11, 62F15, 86A22

1 Introduction

The increasing demand on electric energy and a global concern on fossil-fuel emission levels encourage the development of renewable-based grid-connected systems. The optimization on the energy production requires continuous improvements in energy sources, conversion stages, and transport networks.

In this context, distributed generation systems (DGS) have been widely used in power distribution, and recently have been proposed to integrate different renewable energy sources contributing to generate green energy with lower transport losses. The use of distributed configuration has different advantages, such as the minimization of harmonics and electromagnetic interferences (EMI), redundancy, reliability and

F. Flores-Bahamonde, H. Valderrama-Blavi, J. A. Barrado Rodrigo, J. M. Bosque and A. Leon-Masich:
DEEEA, Universitat Rovira i Virgili, Av. Països Catalans 26, Tarragona, Spain, emails:
freddy.flores@urv.cat, hugo.valderrama@urv.cat.

De Gruyter Oldenbourg, ASSD – Advances in Systems, Signals and Devices, Volume 3, 2017, pp. 19–38.
DOI 10.1515/9783110448412-002

standardization of the power architectures [1, 2]. The connection of different renewable sources to a regulated DC-bus can include storage devices to counteract the randomness of the energy produced,allowing the injection of energy into the grid according to a given profile [3].

Figure 1 illustrates a distributed generation system under development. This is based on a variable DC-bus (270-370 V) ascore of the system. The plant is tied to the grid in a single connection point (PCC) by a 6 kW inverter. All the generators, storage elements, and loads are connected to the DC-bus through an adaptor circuit [4].

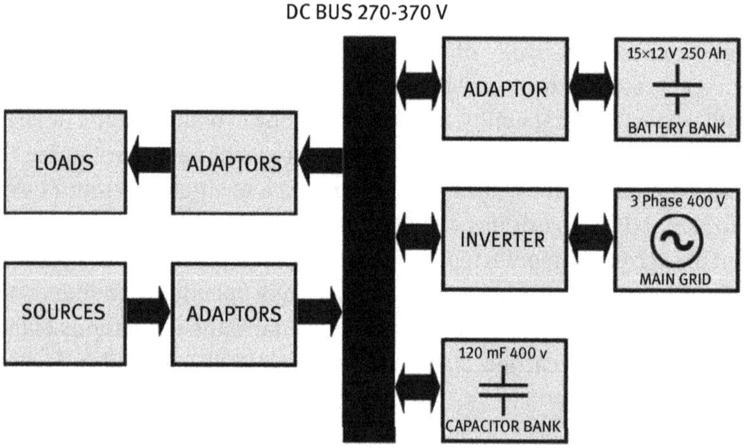

DC BUS 270-370 V

Fig. 1. DC-Microgrid.

The design of power converters with power factor correction (PFC) capability, high efficiency and low cost is still an important research subject in power conversion systems. In this context, the integration of wind power sources to the distributed generation system requires the study of AC/DC converters to optimize the energy injected to a high voltage regulated DC-bus.

In the literature, different topologies applied to the conversion process of wind turbine can be found. For instance, the six-switch rectifier [5, 6], used as a first stage in back-to-back two-level configuration, and the Vienna [7] rectifier used in NPC three-level structures [8], are commonly power stages used for AC/DC conversion in high power wind systems. Nevertheless, other structures for low power application, such as the single-phase AC/DC converter composed of an uncontrolled rectifier and a boost converter stage, are commonly used also in some applications [9]. Moreover, some authors have proposed to use three single-phase rectifiers, one per phase, as a suitable solution. Thus, the power is split in three stages offering modularity and easy implementation [10, 11].

The main idea behind this study is to evaluate the performance, efficiency, and complexity of the different three-phase PFC rectifiers in order to optimize the efficiency of the system maximizing the energy delivered by the generator. For that, repeatable experiments emulating different weather conditions, resulting in different power levels, must be carried out

According to that, a wind system simulator is proposed to evaluate AC/DC converters for connecting a wind generator to a distributed system. Besides, the simulator tool can be also used to evaluate the performance of different MPPT algorithms.

The whole system is illustrated in Fig. 2. The system proposed is divided in two different parts. The first one shows the wind simulator that consist of a wind work-bench composed of a wind generator coupled mechanically to an induction motor, and driven by a variable frequency drive (VFD). The main idea is to control the wind workbench by means of Labview, in that way that the work-bench behaves as a wind generator for a given certain wind speed profile turbine power coefficient $C_p(\lambda, \beta)$.

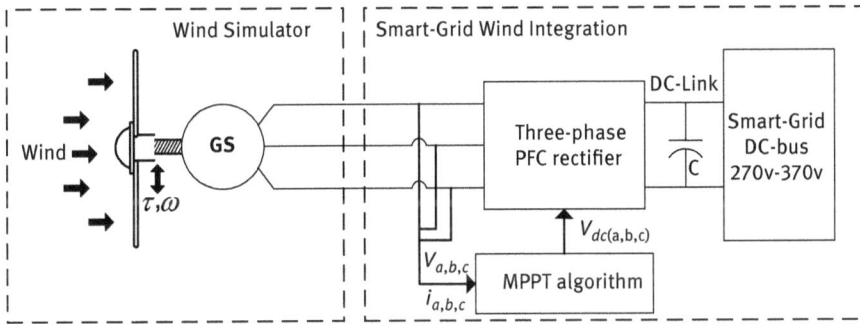

Fig. 2. Wind Generator Simulator Block Diagram.

The second part of the work consists in testing the simulator, connecting the wind generator to an emulated DC-bus using different three-phase rectifier. Besides, a MPPT algorithm will be implemented to extract the maximum power of the wind generator.

The remaining part of the paper is organized as follows. In Section 2, mathematical details concerning the wind system simulator are given. Next, in Section 3, the implementation of the wind simulatoris explained. In section 4, different rectifiers are evaluated. First, the circuits of the different three-phase rectifiers are given. Next, an example of a classical MPPT algorithm is programed in a PIC controller to drive the rectifier. In Section 5, some preliminary experimental results are given, and finally in Section 6, some conclusions are summarized.

2 Wind generator simulator

To design a wind simulator is necessary to consider important aspects, such as wind energy power fluctuations, transient effects, and rotor performance. The importance of these aspects, come from the need to determine how much energy can be extracted from the wind.

In this context, it is well known that the aerodynamic system of a wind turbine is composed mainly of the rotor blades. The air speed v is reduced by the rotor, and the kinetic energy absorbed from the air E_{kin} is transformed in mechanical power P_{mec}.

By [12–14], the kinetic energy from a given cross section of air can be calculated in (1), where ρ is the air density, A the area of the rotor blades, and v_{wind} the wind speed.

$$\int_0^t dE_{kin} = \frac{1}{2}\rho A(v_{wind}t)v_{wind}^2 \rightarrow P_{wind} = \frac{dE_{kin}}{dt} \tag{1}$$

Applying the derivative to the energy expression shown in (1), the power available in the wind can be easily deduced:

$$P_{wind} = \frac{1}{2}\rho A v_{wind}^3 \tag{2}$$

For an air stream, the mechanical power extracted from the wind by an energy transducer is expressed as the difference between the power available in the wind, before and after the transducer, as shown in (3), where v_1 represents the air stream speed before, and v_2 represents the air stream speed after the energy transducer:

$$P_{mec} = \frac{1}{4}\rho A(v_1^2 - v_2^2)(v_1 + v_2) \tag{3}$$

Equation (3) can be expressed as (4), where C_p is the ratio of the mechanical power extracted by the converter, called power coefficient:

$$P_{mec} = C_p(\lambda, \beta)P_{wind} \tag{4}$$

This coefficient is a nonlinear expression that depends on the tip-speed ratio λ (5), and the pitch angle β. The power coefficient represents the power percentage that is actually converted in mechanical power. This concept was introduced by Albert Betz in 1920, who also demonstrated the existence of a physical upper-limit for that value.

$$\lambda = \frac{\omega_m R}{v}, \quad C_p < 0.59 \tag{5}$$

The mathematical expression of the rotor coefficient is the key for wind turbine simulator. In this context different graphical and numerical expressions for power and torque coefficient aims to characterize vertical-axis or horizontal-axis wind rotors [12–14].

For this reason, different studies have been developed to obtain an approximation of $C_p(\lambda, \beta)$, as a result of experimental tests [15], mathematically [16] or analytically from fluid dynamics theory applied to a given rotor type. In fact, the expression of $C_p(\lambda, \beta)$ (6) includes several coefficients (c_1 to c_6), that must be calculated for each real turbine, and depend on some aspects, such as the rotor design, the number and shape of blades, the weight, stiffness and son on. On the other hand, such rotor design aspects cause conversion losses, and therefore the $C_p(\lambda, \beta)$ in a real turbine will be always lower than shown in (5).

To test the wind simulator, a certain expression of $C_p(\lambda, \beta)$ is required. As the generator used in our system come from a horizontal turbine with three blades, we propose the expression (6) with the coefficients defined as $c_1 = 0.5$, $c_2 = 116$, $c_3 = 0.5$, $c_4 = 0$, $c_5 = 5$ and $c_6 = 21$ [9]:

$$C_p = c_1 \left(\frac{c_2}{\lambda_i} - c_3 \beta - c_4 \beta^x - c_5 \right) \exp\left(-\frac{c_6}{\lambda_i} \right) \tag{6}$$

where:

$$\lambda_i = \frac{1}{\dfrac{1}{\lambda + 0.08} - \dfrac{0.035}{\beta^3 + 1}} \tag{7}$$

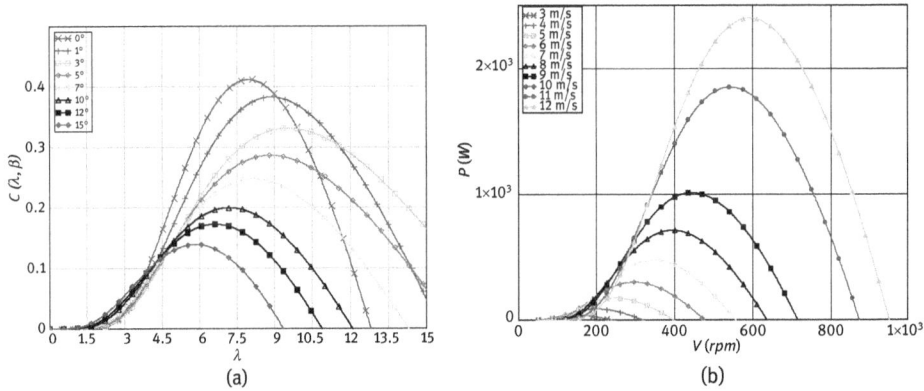

Fig. 3. (a): C_p versus λ for different pitch angles β. (b): P_{mec} in terms of rotor speed and wind speed.

Figure 3a shows the characteristic curves for the power coefficient $C_p(\lambda, \beta)$ for different values of λ and β. These curves are obtained by substituting (7) in (6) and using the length of the rotor blades of the generator ($R = 1.35$ m). In the same way, from (4) and (6) is possible to obtain the waveforms for the power extracted in function of the rotor speed in rpm for different wind velocities, between 3m/s and 12m/s, observed in Fig. 3b. Wind turbine aerodynamic equations show that the power available in the wind P_{wind} depends on the cube of the wind speed. By other hand, the power really extracted from the wind P_{mec} depends also on the rotor angular speed ω, and the pitch angle of the rotor blade β. Thus, the wind simulator has to control the work-bench to obtain an output power from the generator following the curves depicted in Fig. 3b.

3 Wind simulator implementation

In this section the implementation of the wind simulator is described, using equations (6) and (7). Figure 4 illustrate the workbench that consist of a permanent magnet synchronous machine, from a horizontal axis turbine, that is mechanically coupled to an induction motor, and driven by a commercial variable frequency drive (VFD). The main idea of the wind simulator is that using a standard scalar open-loop torque control the VFD drives the induction machine emulating the wind turbine mechanical characteristics, i.e. emulate the turbine shaft [17].

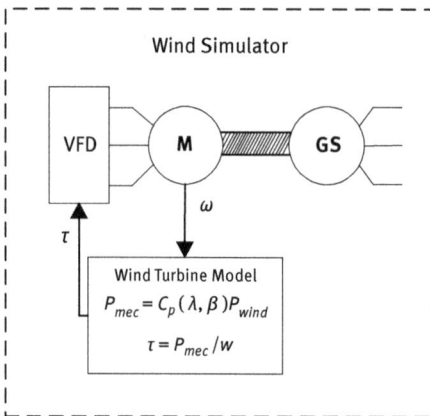

Fig. 4. Wind work-bench emulator.

The simulator will have three input variables: the wind speed v, the rotor angular speed ω, and the pitch angle of the rotor blades β. The flow-chart in Fig. 5 shows how the wind speed, and the rotor angular speed are used to calculate the power available in the wind P_{wind}, the power coefficient C_p, and finally a torque reference for the induction motor. Therefore, using the graphic platform Lab-view, the torque reference τ is calculated according to Fig. 5, and is sent to the VFD, previously programmed using sensor-less Vectorial Control setting.

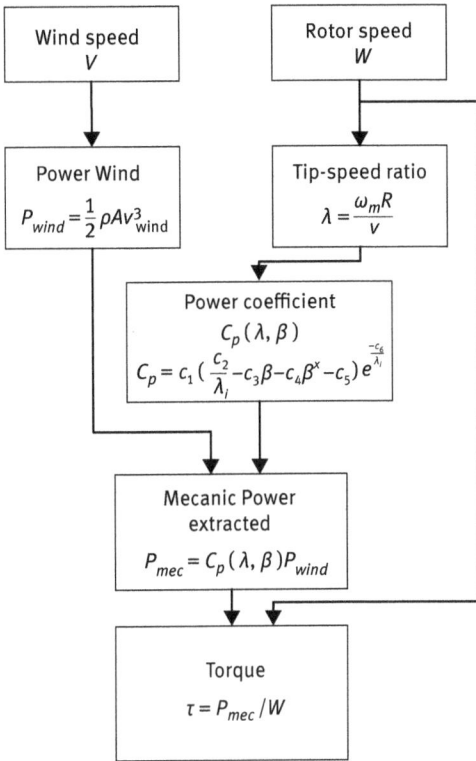

Fig. 5. Torque calculation Flowchart.

To initialize the program, the first step is the introduction of the input variables as the wind speed and the rotor blades pitch angle, and then, the calculation of the rotor speed that is measured by an encoder continuously. The communication between the variable frequency drive and the encoder with the computer is realized by the PCI-6024E data acquisition card from National Instruments.

The flowchart shown in Fig. 6 illustrates the program design in Lab-view, where, the first step is to initialize the input variables. After, a given profile of wind can be programmed and charged in the code. This profile can be previously generated by statistic methods according to a given climate profile in a geographic area. Once the profile is charged, a sub-routine is triggered for calculating the rotor speed. In this sub-routine, each 20 ms, the encoder pulses are counted, and the rotor speed is calculated and filtered by the software to eliminate measurement noise.

On the other hand, once the rotor speed ω is calculated, different sub-routines are executed in a concurrent form. According to Fig. 6, from the left side, the first sub-routine calculates the torque reference as is described in Fig. 5, and sent it to the VFD through the DAQ assistant (digital analog toolbox of Labview). The next sub-routine is designed to detect any changes in the wind speed. If any change occurs, the simulator calculates and draws the new power curves as shown in Fig. 7.

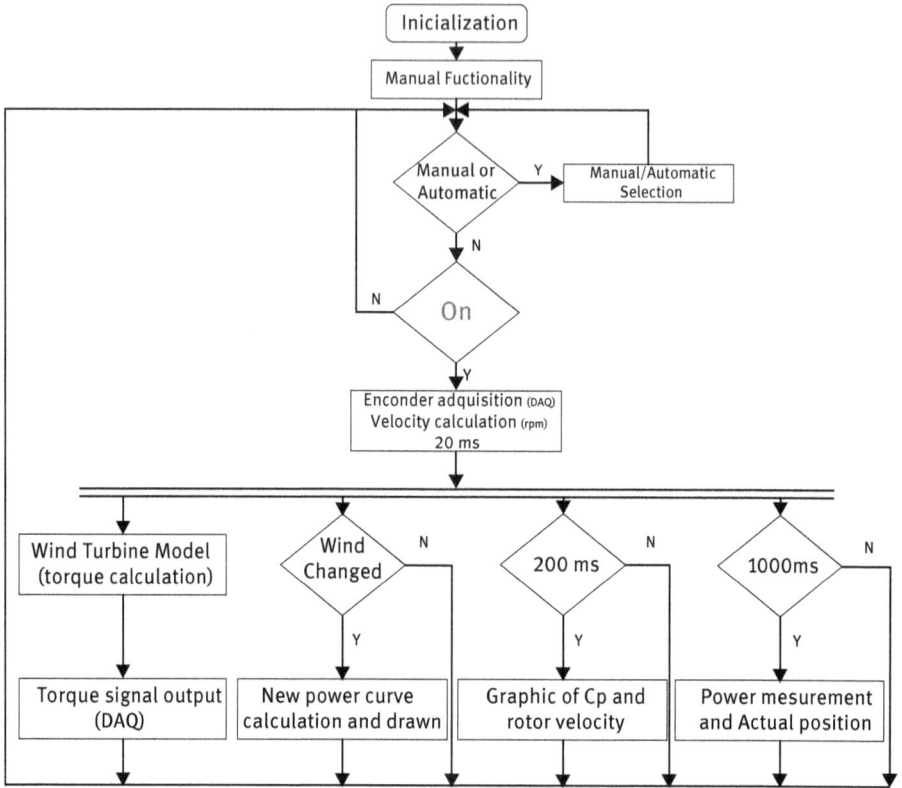

Fig. 6. Flowchart of the Labview-programmed code.

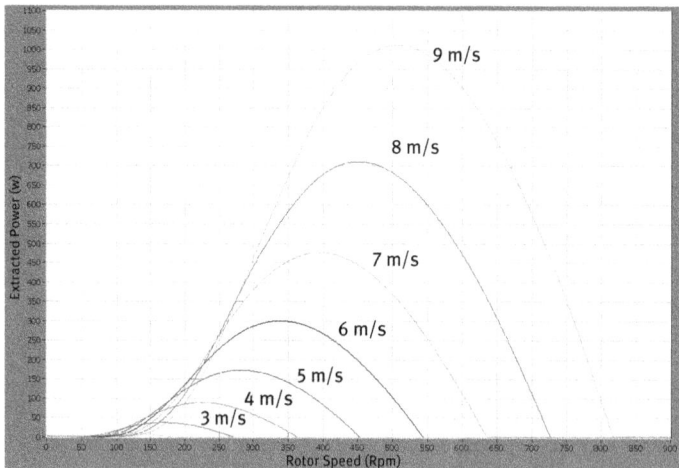

Fig. 7. Mechanical power curves obtained from Labview.

The next subroutine shows the information of different variables to the user, for instance: the rotor speed, power available, mechanical power and the power coefficient. The last routine is devoted to refresh the power extracted from the wind, and measured in the output of the generator. This information is graphically shown over the power curve drawn for a given wind speed. This is considered important information, because shows the actual point where the system is working, and also it allows to observe graphically when the maximum power point is achieved.

4 AC-DC conversion and MPPT algorithm

To evaluate the wind simulator implemented in the previous section, we connect different three-phase rectifiers with power factor correction (PFC) at the output of the synchronous generator. This section, is divided in two parts. The first is dedicated to the explanation of the topologies used to extract the power of the wind turbine. The second part is devoted to a MPPT algorithm implemented to extract the maximum power of the wind turbine.

4.1 PFC Boost-Based Rectifiers

To evaluate the emulator, three different rectifier topologies will be connected to the output of the wind generator. All the energy delivered is injected to a high-voltage DC-bus emulated by a power supply, as is depicted in Fig. 8. Figure 9 illustrates the rectifiers. A rectifier composed of three modular single-phase rectifiers is shown in Fig. 9a, the six-switch rectifier in Fig. 9b, and the Vienna rectifier in Fig. 9c. The three rectifiers are controlled to behave as a loss-free resistor (LFR) by means of sliding control mode [18].

This means that, imposing a sliding surface to the inductor current, the three-phase rectifier is seen by the generator as three equal resistances, one per phase, obtaining a good power factor correction. As the currents have sinusoidal waveforms, the winding losses are reduced and the torque vibrations are eliminated.

The switching surface, shown in (8), must induce the sliding motions in each phase of the rectifier, where g (expressed in Ω^{-1}) is the input conductance of each phase, and V_j is the input voltage of that j^{th} phase. At permanent regime is reached $S(x) = 0$, and then $i_{Lj} = gV_j$. Consequently the current will track its respective input voltage [10, 11, 18] (for $j = A, B, C$):

$$S_j(x) = i_{Lj} - gV_j = 0 \tag{8}$$

Then we can calculate the power delivered by each phase as:

$$P_j(t) = V_j i_j \tag{9}$$

Solving (9), and assuming permanent regimen conditions is applied (8), we obtain

$$\begin{cases} P_A(t) &= gV_m^2 \sin^2 \omega t \\ P_B(t) &= gV_m^2 \sin^2 \left(\omega t + \frac{2\pi}{3}\right) \\ P_C(t) &= gV_m^2 \sin^2 \left(\omega t - \frac{2\pi}{3}\right) \end{cases} \tag{10}$$

where V_m is the peak voltage of the sinus wave. On the other hand, an assuming an ideal converter without losses, the output power can be expressed as

$$P_0 = P_A + P_B + P_C = \frac{3}{2}gV_m^2 \tag{11}$$

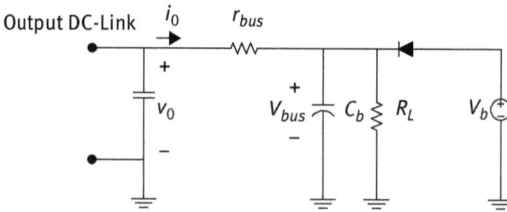

Fig. 8. Emulated bus-DC.

Applying some manipulations is easy to deduce that the current injected into the DC-bus can be expressed as:

$$I_{bus} = \frac{P_0}{V_{bus}} = \frac{3gV_m^2}{2V_{bus}} \tag{12}$$

Equation (11) shows that the surface expressed in (8) performs simultaneously a power factor correction for each phase, allowing also, a direct control of the power extracted from the wind turbine. This power can be controlled adjusting the input conductance (g) of each phase. This conductance that now is controlled externally by an arbitrary voltage reference, in the future can be controlled by the MPPT algorithm.

4.2 MPPT algorithm

Although the power extracted from the wind depends on the cube of the wind speed, and the power coefficient $C_p(\lambda, \beta)$ depends on both, rotor speed and wind speed, it is not convenient to develop a MPPT algorithm sensing the wind speed, as such sensors are slow and imprecise [14]. For this reason, MPPT algorithms normally are based on electrical variables, for example, the rotor power.

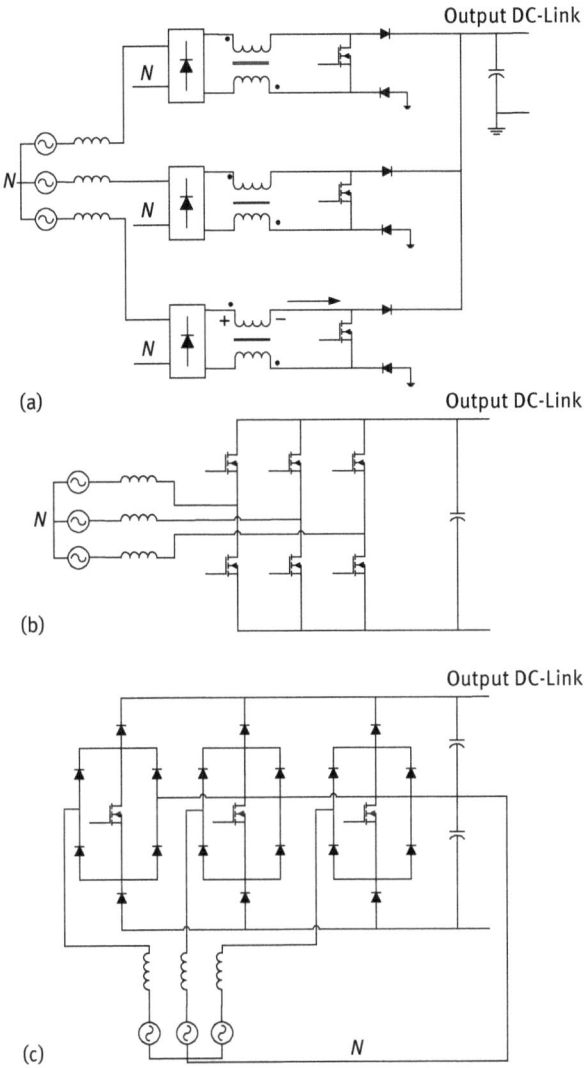

Fig. 9. (a) Modular single-phase rectifier. (b) Six-switch rectifier. (c) Vienna rectifier.

According to [19, 20], if the curves depicted in Fig. 10a are taken into account, it can be deduced that the maximum power can be found by means of (8). The algorithm operation for different wind speed can be observed in Fig. 10b, where $v_2 > v_3 > v_1$.

How is demonstrated in [20], applying some mathematical manipulations, expression (13) becomes (14), where V_{gen} is the output turbine voltage, ω the rotor speed, and ω_e the voltage phase angle speed generated by the turbine:

$$\frac{dP_{mec}}{d\omega} = 0 \qquad (13)$$

(a)

(b)

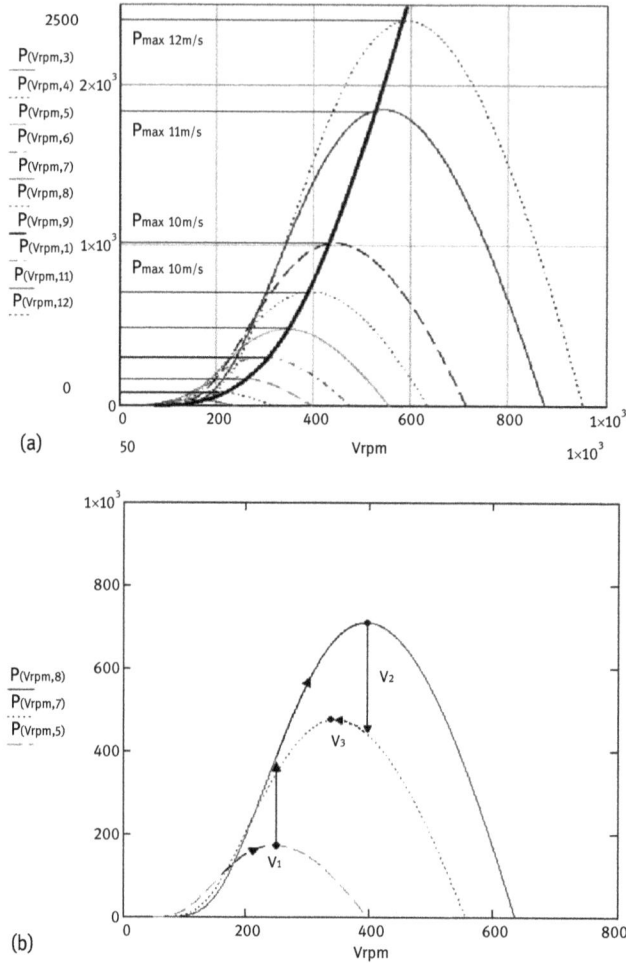

Fig. 10. (a): Maximum power points on the wind power curve. (b): MPPT algorithm performance.

where:

$$\frac{dP_{mec}}{d\omega} = \frac{dP_{mec}}{dV_{gen}} \times \frac{dV_{gen}}{d\omega_e} \times \frac{d\omega_e}{d\omega} \quad (14)$$

Through (13) and (14), (15) is obtained and is possible find the maximum power controlling the output power of the wind turbine. Therefore, the maximum power extracted is achieved controlling V_{gen}, thus avoiding the use of mechanical variables as ω_e, and obtaining a better performance, specifically in precision terms. Thus:

$$\frac{dP_{mec}}{d\omega} = 0 \rightarrow \frac{dP_{mec}}{dV_{gen}} = 0 \quad (15)$$

The algorithm used in this work is shown in the flowchart given in Fig. 11. The algorithm is initialized with an arbitrary voltage reference V_{dc}, where V_{dc} is a proportional value of the conductance given by $V_{dc} = kg$. Next, the measurement of the voltage and current in the generator output is realized by the controller, and a cyclic process begins until the maximum power point is reached.

The expression V_{dc} represents the DC-voltage level introduced at the rectifier topology to vary the conductance of each phase. Therefore, and increment or decrement of V_{dc} produces directly and increment or decrement of the rectifier conductance.

Note that, the output voltage is imposed by the DC-bus, thus,the currents and voltage signals using by the algorithm are measured at the rectifier input and have a rectified sinusoidal waveform.Besides, the voltage of the generator has a range of frequency which varies between $20 - 80$ Hz and consequently between $40 - 160$ Hz for the rectified waveforms.Whatever be the frequency then umbers of samples have to be sufficiently high to calculate the mean value of each semi-cycle. On the other

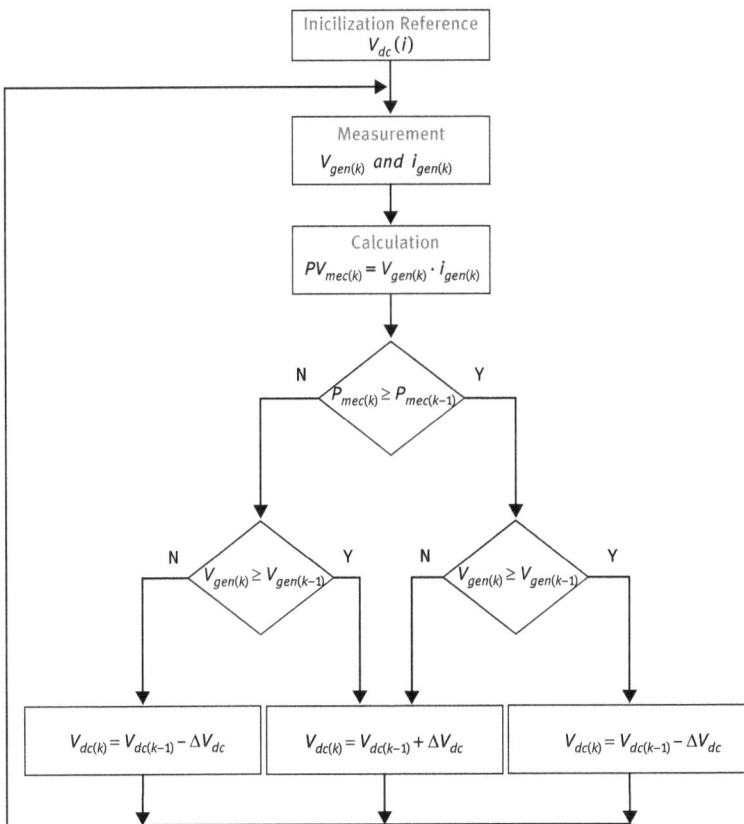

Fig. 11. Classical MPPT algorithm flow-chart.

hand, as well known, it is mandatory to realize the conductance changes, i.e. power changes, in the end of each semi-cycle.

Therefore, using a dsPIC30F2020 controller, the algorithm described in Fig. 11 was implemented, where 85 samples of each-semi cycle of i_{gen} and V_{gen} are accumulated to calculate the mean value. Moreover, taking into account that the mechanical system have a time constant to reach the steady state, the power calculated used by the algorithm $P_{mek(k)}$ is the result of averaging the last 5 power samples, this means, that, the averaging of each 5 semi-cycles of the signal. The number of samples used to calculate the power value $P_{mek(k)}$ can be adapted by the user to adjust the delay needed for stabilize the MPPT algorithm [20, 21]. Finally, once the algorithm has found the MPP the system oscillate around this point.

5 Experimental results

To verify the wind simulator design, and prove its functionality, different experiments have been realized to evaluate its performance.

The main idea of the wind simulator is to study the performance of both, PFC rectifier converters, and the MPPT algorithms. Therefore, the first experiment is the connection of the rectifiers shown in Fig. 9. The converters are connected to the output of the wind turbine, and also connected to a dc-bus composed of a capacitors bank, a power supply and a load, Fig. 8.

The picture shown in Fig. 12 depicts the wind work-bench, whereas the Fig. 13a, 13b and 13c shown the steady state waveforms for current of the rectifiers of the

Fig. 12. Work-bench connected to a three-phase rectifier.

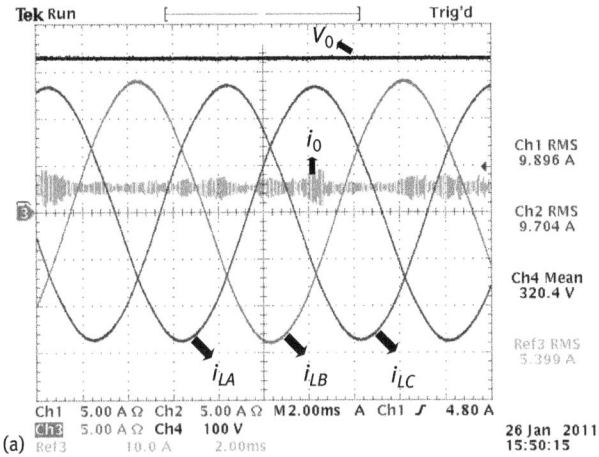

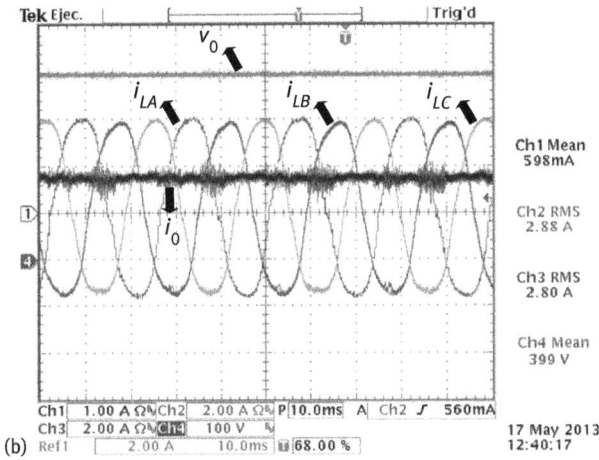

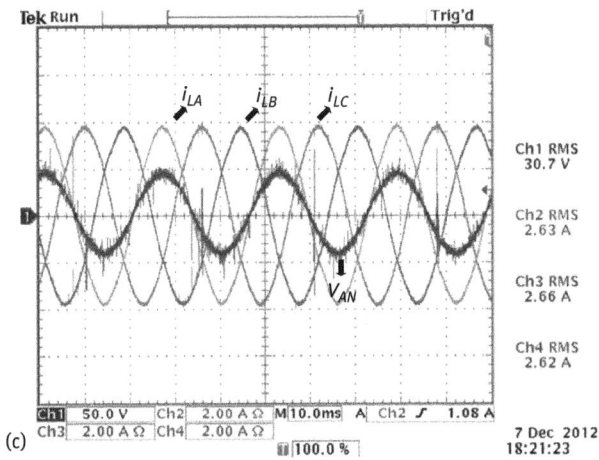

Fig. 13. (a): Steady Bus experiment for modular 3-phase rectifier. (b): Steady Bus experiment for Vienna-switch rectifier. (c): Steady Bus experiment for six-switch rectifier.

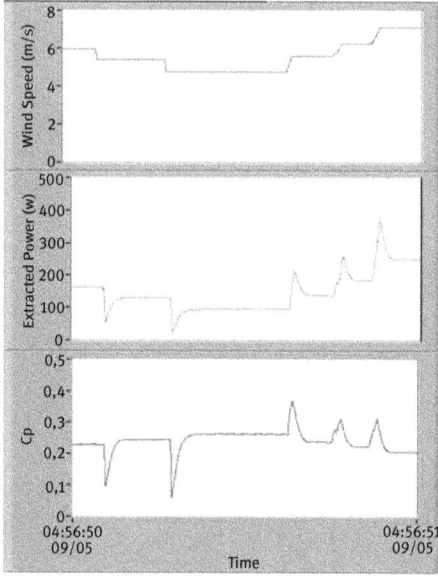

(a)

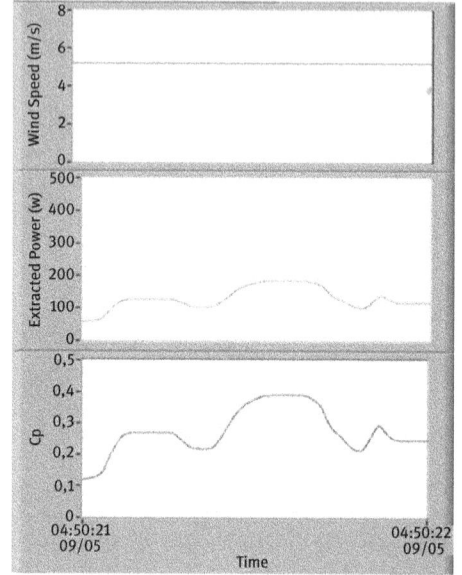

(b)

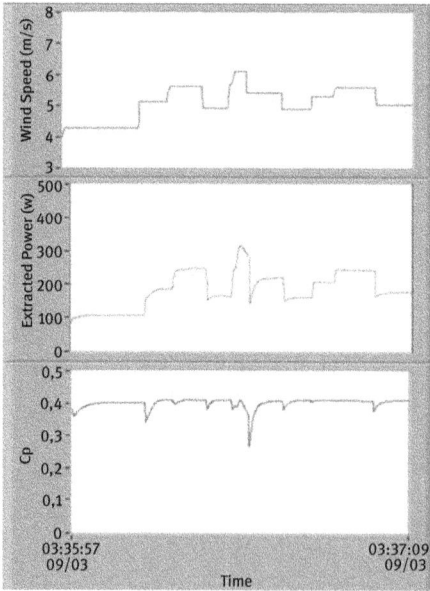

(c)

Fig. 14. (a): Constant conductance and variable wind speed. (b): Constant wind speed and variable conductance. (c): Given profile wind speed and MPPT algorithm.

Fig. 9a, 9b and 9c respectively. All the rectifiers are connected to a approximately 300 V dc-bus.

As has been demonstrated in the theoretical analysis, by means of adjusting the input conductance the power extracted from the wind turbine can be controlled. In addition, the rectifier conductance can be modified manually with the aid of an external DC power supply. Then, if a DC-voltage excursion is carried out, it is possible to explore the power curves shown in Fig. 7.

The next experiment is shown in Fig. 14a. Keeping constant the conductance values, a wind profile is programmed as depicted in the upper part of that figure. The corresponding extracted power is depicted in the central part of the figure, and finally, in the inferior part of the figure, the power coefficient is shown.

A similar experiment is realized and depicted in Fig. 14b. Now, the conductance is randomly varied for a given wind speed (5 m/s). This experiment demonstrates that adjusting the rectifier conductance is possible to control the amount of power extracted for a certain given wind speed.

Finally, the experiment shown in Fig. 14c is the most important of all.Here the manual conductance adjustment is replaced by the MPPT controller explained in Section III.

In the upper part of the figure a random wind profile is given. Next, in the central part of such figure, we can observe the extracted power, and finally the inferior part the power coefficient C_p is shown. As we can realize, the power coefficient is always around the maximum point, with the exception associated to the wind-speed transients.

This last experiment demonstrates the correct behavior of the MPPT algorithm, proving the importance of the wind simulator as a platform tool for repeatable experiments.

6 Conclusions

The developed work in this project is a contribution for the study of renewable energy systems, specifically for the wind energy. The main idea is to design a simulation platform to realize repeatable experiments to study the performance of both, the conversion chain converters, and the MPPT algorithm.

After programming a specific $C_p(\lambda, \beta)$ curve, the user by means of the Labview graphic interface, can introduce a wind profile and then observe the $P - \omega$ curve, the power extracted, and the power coefficient $C_p(\lambda, \beta)$ of the wind turbine. Besides, measuring the power output is possible to situate and observe the actual generator operation point in the power curve $P - \omega$.

By other hand, the behavior of the simulator has been verified connecting a three-phase rectifier at the output of the wind turbine, developing also a MPPT algorithm to extract the maximum power point or equivalently to operate at a constant value of the power coefficient $C_p(\lambda, \beta)$.

The wind simulator will be used to compare the behavior of different rectifier topologies with power factor correction, and to evaluate the global conversion chain.

Acknowledgement: This work has been partially sponsored by the Spanish Ministry of Research and Science, under grants: DPI2009-14713-C03-02, and Consolider RUE CSD2009-00046.

Bibliography

[1]   S. Luo and I. Batarseh. A review of distributed power systems. Part I: DC distributed power system. *IEEE Aerospace Electronic System Magasine*, 21(6-2):5–14, June 2006.
[2]   K. Kurohane, T. Senjyu, A. Yona, N. Urasaki, E.B. Muhando and T. Funabashi. A high quality power supply system with DC smart grid. *Transmission and Distribution Conf. and Exposition*, 2010 IEEE PES, :1–6, April 19–22, 2010.
[3]   C. Millais and L. Colasimone. *Large Scale Integration of Wind Energy in the European Power Supply: Analysis Issues and Recommendations*. Report of the European Wind Energy Association (EWEA), 2005.
[4]   H. Valderrama-Blavi, J.M. Bosque, F. Guinjoan, L. Marroyo and L. Martinez-Salamero. Power Adaptor Device for Domestic DC Microgrids Based on Commercial MPPT Inverters. *IEEE Trans. on Industrial Electronics*, 60(3):1191–1203, March 2013.
[5]   H. Zhang and B. Tan. Simulation Research on Three-Phase Six-Switch PWM Rectifier with One Cycle Control. 2nd *Int. Conf. on Intelligent Computation Technology and Automation*, ICICTA '09., 2:244–247, October 10-11, 2009.
[6]   M. Liserre, R. Cardenas, M. Molinas and J. Rodriguez. Overview of Multi-MW Wind Turbines and Wind Parks. *IEEE Trans. on Industrial Electronics*, 58(4):1081–1095, April 2011.
[7]   Hao Chen and D.C. Aliprantis. Analysis of Squirrel-Cage Induction Generator with Vienna Rectifier for Wind Energy Conversion System. *IEEE Trans. on Energy Conversion*, 26(3):967–975, September 2011.
[8]   E.J. Bueno, S. Cobreces, F.J. Rodriguez, A. Hernandez and F. Espinosa. Design of a Back-to-Back NPC Converter Interface for Wind Turbines with Squirrel-Cage Induction Generator. *IEEE Trans. on Energy Conversion*, 23(3):932–945, September 2008.
[9]   Y.Y. Xia, J.E. Fletcher, S.J. Finney, K.H. Ahmed and B.W. Williams. Torque ripple analysis and reduction for wind energy conversion systems using uncontrolled rectifier and boost converter. *Renewable Power Generation*, IET, 5(5):377–386, September 2011.
[10] F. Flores-Bahamonde, H. Valderrama-Blavi, J.M. Bosque and L.Martinez-Salamero. Modular-based PFC for low power three-phase wind generator. 7th *Int. Conf.-Workshop Compatibility and Power Electronics* (CPE), :125–130, June 1-3, 2011.
[11] F. Flores-Bahamonde, H. Valderrama-Blavi, J.M. Bosque-Moncusi, L. Martinez-Salamero, A. Leon-Masich and J.A. Barrado. Single-Phase PFC for Three-Phase Wind Generator, a Modular Approach. 915147: *Przeglad Elektrotechniczny*. 88(1):56–60.
[12] S. Heier. *Grid Integration of Wind Energy Conversion Systems*. Wiley, 1998.
[13] Z. Lubosny. *Wind Turbine Operation in Electric Power Systems*. Berlin, Germany: Springer, 2003.

[14] E. Hau. *Wind Turbines: Fundamentals, Technologies, Application, Economics*. 2nd edition. Springer 2005.

[15] Hsiang-Chun Lu and Le-Ren Chang-Chien. Use of wind turbine emulator for the WECS development. *Int. Power Electronics Conf.* (IPEC), 3188–3195, June 21–24, 2010.

[16] A. Monroy and L. Alvarez-Icaza. Real-time identification of wind turbine rotor power coefficient. *45th IEEE Conf. Decision and Control*, 3690–3695, December 13–15, 2006.

[17] R. Teodorescu and F. Blaabjerg. Flexible control of small wind turbines with grid failure detection operating in stand-alone and grid-connected mode. *IEEE Trans. on Power Electronics* 19(5):1323–1332, September 2004.

[18] A. Cid-Pastor, L. Martinez-Salamero, A. El Aroudi, R. Giral, J. Calvente and R. Leyva. Synthesis of loss-free resistors based on sliding-mode control and its applications in power processing. *Control Engineering Practice*, 21(5):689–699, May 2013.

[19] D.A. Caixeta, G.C. Guimaraes, M.L.R. Chaves, J.C. Oliveira and A.F. Bonelli. Maximization of variable speed wind turbine power including the inertia effect. *11th Int. Conf. on Electrical Power Quality and Utilisation* (EPQU), :1–6, October 17–19, 2011.

[20] E. Koutroulis and K. Kalaitzakis. Design of a maximum power tracking system for wind-energy-conversion applications. *IEEE Trans. on Industrial Electronics*, 53(2):486–494, April 2006.

[21] R. Leyva, C. Alonso, I. Queinnec, A. Cid-Pastor, D. Lagrange and L. Martinez-Salamero. MPPT of photovoltaic systems using extremum – seeking control. *IEEE Trans. on Aerospace and Electronic Systems*, 42(1):249–258, January 2006.

Biographies

Freddy Flores-Bahamonde was born in Osorno, Chile, in 1983. He received the B.S in electronical engineering, in 2007 from Universidad Tecnològica Metropolitana de Chile in Santiago de Chile. In 2009 received the M.S degree from Universitat Rovira i Virgili in Tarragona, Spain. From 2010 he is a member of the GAEI research group (Universitat Rovira i Virgili) on Industrial and Automatic Control whose main research fields are power electronics, especifically AC/DC converters for wind energy integration. He is currently working towards the Ph.D degree at Universitat Rovira i Virgili, from Tarragona, Spain.

Hugo Valderrama-Blavi received the Ingeniero and Ph.D. degrees from the Universitat Politècnica de Catalunya, Barcelona, Spain, in 1995 and 2001, respectively. He is currently an Associate Professor in the Departament d'Enginyeria Electrònica, Elèctrica I Automàtica, Universitat Rovira i Virgili, Tarragona, Spain. During the academic year 2001-2002, he was a Visiting Scholar at Laboratoire d'Automatique et Analyse des Systèmes, Centre National de la Recherche Scientifique, Toulouse, France. His cur-rent research interests are power electronics, renewable energies, silicon carbide devices, and nonlinear control.

José Antonio Barrado Rodrigo received the electronic engineering degree from the Universitat de Barcelona (UB), Spain, in 2000 and the Ph.D. degree in automatic control from Universitat Politècnica de Catalunya (UPC), Barcelona, Spain, in 2008. He is currently an Associate Professor with the Department of Electronic Engineering and Automatic Control of the Universitat Rovira i Virgili (URV), Tarragona, Spain. His research interests include analysis, modeling and control of electric generators and power converters applied to renewable energy systems.

Josep M. Bosque received the Enginyer Tècnic Industrial en Electrònica Industrial degree and the Enginyer en Electrònica master degree from the Universitat Rovira i Virgili (URV), Tarragona, pain, in 2005 and 2009, respectively, where he is currently working toward the Ph.D. degree. Since 004, he has been a Research Technician with the Automatics and Industrial Electronics Group, URV. is research interests are power electronics and renewable energy.

Antonio Leon-Masich was born in Lleida, Spain, in 1986. He received the B.S and the M.S degrees in electronical engineering, in 2009 and 2011 respectively from Universitat Rovira i Virgili in Tarragona, Spain. From 2009 he is a member of the GAEI research group (Universitat Rovira i Virgili) on Industrial and Automatic Control whose main research fields are high-gain converters and high-voltage converters for electronic ballast using Silicon carbide devices. He is currently working towards the Ph.D degree at Universitat Rovira i Virgili, from Tarragona, Spain.

A. Farah, T. Guesmi, H. Hadj Abdallah and A. Ouali

On line improvement of power system dynamic stability using ANFIS and NSGA II algorithms

Abstract: This paper investigates the use of an adaptive power system stabilizer (PSS) for improving the dynamic stability of a power system. Adaptive network based fuzzy inference systems (ANFIS) and the second version of non-dominated sorting genetic algorithms (NSGA-II) is employed to select the optimal parameters of the controller for different loading conditions. Firstly genetic algorithms are used to tune stabilizer parameters on a wide range of loading conditions to create a data base. Two eigenvalue-based objective functions are considered to place the closed-loop system eigenvalues in the D-shape sector. Then, the relationship between these operating points and the corresponding stabilizer parameters is learned by the ANFIS. The proposed stabilizer has been tested by performing non linear simulations and eigenvalue analysis using single machine infinite bus (SMIB) model. The results show the effectiveness and the robustness of the proposed stabilizer to provide efficient damping in real-time.

Keywords: Power system stability, PSS, NSGA-II, ANFIS.

Mathematics Subject Classification 2010: 65C05, 62M20, 93E11, 62F15, 86A22

1 Introduction

The stability of power systems is a challenging problem in electric system operation. This problem can be defined as the ability of the power system to return to normal operating state when subject to disturbance [1]. Currently, power system stabilizers (PSS) are routinely used to improve the dynamic stability of power systems by controlling the excitation of the generator. In general, the function of the PSS is to produce component of electrical torque in phase with the rotor speed deviations.

The conventional PSS (CPSS) is widely presented as a lead-lag compensator [2, 4]. In [1, 5], the authors have presented a comprehensive study of the effects of the parameters of this lead-lag model. Recently, considerable researches have been focused on the designing and using of new adequate damping sources [1–6]. The robust design of the PSS is evaluated by several operating conditions [3].

A. Farah, T. Guesmi, H. Hadj Abdallah and A. Ouali: University of Sfax, National Engineering School of Sfax-Tunisia, Control & Energy Management Laboratory, email, farah.anouar@gmail.com, tawfik.guesmi@istmt.rnu.tn, hsan.haj@enis.rnu.tn, abderrazak.ouali@enis.rnu.tn.

De Gruyter Oldenbourg, ASSD – Advances in Systems, Signals and Devices, Volume 3, 2017, pp. 39–53.
DOI 10.1515/9783110448412-003

The design problem is formulated as an optimization problem. Multiple conventional methods are used to solve this design problem, such as eigenvalues assignment, mathematical programming, and gradient procedure. These techniques are iterative and consume an important computing time. Also, it may converge to local optimum. To overcome these disadvantages, evolutionary algorithms are becoming the most popular in this type of problem. In this context, a second version of non-dominated sorting genetic algorithms (NSGA-II) [8] is adapted in this paper, to generate the Pareto optimal solutions.

Larsen and Swann [2] have demonstrated that the PSS can be very well tuned to an operating point and provide satisfactory damping over a certain range around the design point. Therefore, the stabilizers cannot guarantee the stability and good performance, if the operating conditions change. So, it is necessary to determine the optimum stabilizer parameters at each operating point, which does not allow an online decision. To overcome these difficulties, novel intelligent techniques based on fuzzy logic and artificial neural network is used in this paper.

This online design is based on multi objective evolutionary algorithms and ANFIS and it is done on two different stages. The first step consists to adjust controller parameters using an improved version of non-dominated sorting genetic algorithms (NSGA-II) [7–9] for a wide range of operating conditions. The closed-loop system eigenvalues should be placed in the D-shape sector. The second step is based on ANFIS training with the collected input-output data pairs which are stocked in the first step. The input data are the operating conditions and the outputs are the controller parameters. To assess the effectiveness of the proposed stabilizers, their performance has been tested on a weakly connected power system. Eigenvalue analysis and nonlinear simulations are carried out.

This paper is organized as follows: Section 2 describes the small signal modeling of the Single Machine Infinite Bus (SMIB) with PSS. The optimization problem has been introduced in section 3. In section 4 the general theory of NSGA-II is presented. The ANFIS is introduced in section 5. Simulations and results are provided and discussed in section 6 and conclusions are given in VII.

2 System modeling

A SMIB system, as shown in Fig. 1, is considered for the damping control design. The equivalent line impedance is $Z = R + jX$. The generator **G** is equipped with a PSS and it has a local load $Y_L = G + jB$. The preliminary system data are given in the appendix. In Fig. 1, V_t and V_0 are the generator terminal and the infinite bus voltages, respectively.

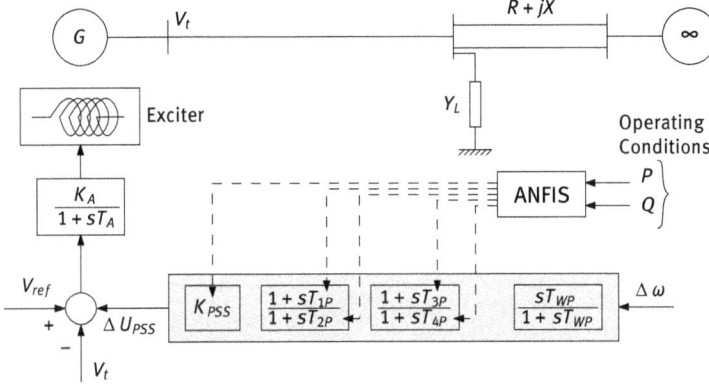

Fig. 1. SMIB system.

2.1 Generator model

In this study, the generator is represented by the third-order model [2, 9, 14, 15], which are, the two motion equations and the generator internal voltage equation.

$$p\delta = \omega_b(\omega - 1) \tag{1}$$

$$p\omega = \frac{P_m - P_e - D(\omega - 1)}{M} \tag{2}$$

$$PE'_q = \frac{E_{fd} - (x_d - x'_d)i_d - E'_q}{T'_{do}} \tag{3}$$

where δ and ω are rotor angle and speed, respectively. ω_b is the base frequency expressed in rad/sec.

In equation (2), P_m and P_e are the mechanical power input and the electrical power output of the generator, respectively. These powers are in per unit (p.u.). D and M are the damping coefficient and inertia constant respectively. For equation (3), E_{fd} and E'_q are the field and the internal voltages, respectively. They are in per unit. i_d is the d-axis armature current x'_d and x_d are the d-axis transient reactance and the d-axis reactance of the generator, respectively. T'_{do} is the open circuit field time constant. The electrical power P_e can be expressed by the d-axis and q-axis components of the internal voltage V_t and the armature current i, as follows [14, 16]:

$$P_e = V_d i_d + V_q i_q \tag{4}$$

$$V_d = X_q i_q \tag{5}$$

$$V_q = E'_q - X'_d i_d \tag{6}$$

$$V_t^2 = V_d^2 + V_q^2 \tag{7}$$

X_q is the q-axis reactance of the generator. Using equations (4–6) the electrical power can be written as:

$$P_e = E'_q i_q + (X_q - X'_d)i_d i_q \tag{8}$$

2.2 Structure of excitation system with PSS

The IEEE type-ST1 excitation system with PSS shown in Fig. 1 is considered in this paper [2, 14], where K_A and T_A are the gain and the time constants of the excitation system, respectively. V_{ref} is the reference voltage and U_{PSS} is the stabilizer signal. The PSS representation consists of a gain K_{PSS}, a washout block with the time constant T_{WP} and two lead-lag blocks.

From the excitation system block, the field voltage E_{fd} can be expressed as:

$$pE_{fd} = \frac{K_A(V_{ref} - V_t + U_{PSS}) - E_{fd}}{T_A} \tag{9}$$

2.3 Philips-Heffron model with PSS

Using previous equations, the linearized form of P_e, E_q and V_t can be expressed by the following equations:

$$\Delta P_e = K_1 \Delta\delta + K_2 \Delta E'_q \tag{10}$$

$$\Delta E_q = K_3 \Delta E'_q + K_4 \Delta\delta \tag{11}$$

$$\Delta V_t = K_5 \Delta\delta + K_6 \Delta E'_q \tag{12}$$

where:

$$K_1 = \frac{\partial P_e}{\partial\delta}, \ K_2 = \frac{\partial P_e}{\partial E'_q}, \ K_3 = \frac{\partial E_q}{\partial E'_q}, \ K_4 = \frac{\partial E_q}{\partial\delta}, \ K_5 = \frac{\partial V_t}{\partial\delta}, \ K_6 = \frac{\partial V_t}{\partial E'_q} \tag{13}$$

These coefficients depend on the operating conditions. Thus, the linearized SMIB model with PSS can be represented in the state-space form by: $\dot{X} = AX + BU$, where:

$$A = \begin{bmatrix} 0 & \omega_0 & 0 & 0 \\ -\dfrac{K_1}{M} & -\dfrac{K_D}{M} & -\dfrac{K_2}{M} & 0 \\ -\dfrac{K_4}{T'_{do}} & 0 & -\dfrac{K_3}{T'_{do}} & \dfrac{1}{T'_{do}} \\ -\dfrac{K_A K_5}{T_A} & 0 & -\dfrac{K_A K_6}{T_A} & -\dfrac{1}{T_A} \end{bmatrix}$$

$$\tag{14}$$

$$B = \begin{bmatrix} 0 \\ 0 \\ 0 \\ \dfrac{K_A}{T_A} \end{bmatrix}, \ X = \begin{bmatrix} \Delta\delta \\ \Delta\omega \\ \Delta E'_q \\ \Delta E_{fd} \end{bmatrix}, \ U = \Delta U_{PSS}$$

3 Optimization problem formulation

In this study, the problem of tuning parameters of the PSS controller that stabilize the system is converted to a multi objective optimization problem. As given in [2, 17], two eigenvalue-based objective functions are considered. The first one consists to shift the closed-loop eigenvalues in to the left-side of the line defined by $\sigma = \sigma_0$, as shown in Fig. 2(a). This function is expressed by J_1 in equation (15). In equation (16), J_2 defines the second objective function. It will place the closed-loop eigenvalues in a wedge-shape sector corresponding to $\xi_{ij} \geq \xi_0$, as shown in Fig. 2(b) As consequence the maximum overshoot is limited.

$$J_1 = \sum_{\sigma_i \geq \sigma_0} (\sigma_0 - \sigma_i) \tag{15}$$

$$J_2 = \sum_{\xi_i \leq \xi_0} (\xi_0 - \xi_i) \tag{16}$$

where σ_i and ξ_i are respectively, real part and damping ratio of the i-th eigenvalue corresponding to an operating point.

Therefore, the coordinated design problem is aimed to minimize simultaneously J_1 and J_2 by respecting the adjustable parameter bounds. So, the closed-loop eigenvalues will be placed in the D-shape sector shown in Fig. 2(c). It is important to recognize that only the unstable or lightly damped electromechanical modes are relocated.

The adjustable parameter bounds are given by the following equations:

$$K_{PSS}^{\min} \leq K_{PSS} \leq K_{PSS}^{\max} \tag{17}$$

$$T_{1P}^{\min} \leq T_{1P} \leq T_{1P}^{\max} \tag{18}$$

$$T_{3P}^{\min} \leq T_{3P} \leq T_{3P}^{\max} \tag{19}$$

The washout time constant T_W and the two lead lag time constants of PSS are usually prespecified, as follows [2]: $T_W = 5s$, $T_{2P} = T_{4P} = 0.1s$.

Other constraint can be added, which is the electromechanical modes frequency limits [17].

$$\omega_{\min} \leq \omega \leq \omega_{\max} \tag{20}$$

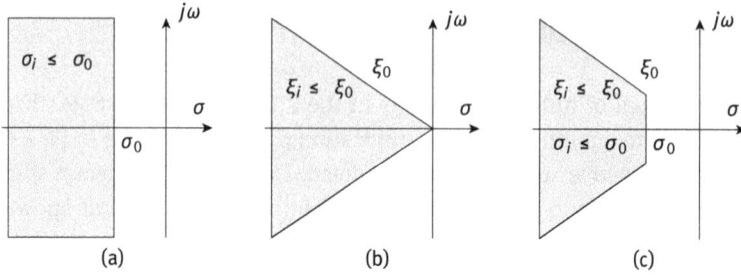

Fig. 2. Region of eigenvalues location for different objective functions.

4 Implementation of the NSGAII approach

Multi objective evolutionary algorithms which use no dominated sorting and sharing, such as, NSGA and NPGA (niched Pareto genetic algorithm) have been criticized for their, high computational complexity, absence of elitism and need for specifying the sharing parameter. Thus, an improved version of NSGA, called NSGAII [7, 8] is proposed in this paper. In this approach, the sharing function approach is replaced with a crowded comparison.

Initially, an offspring population Q_t is created from the parent population P_t at the t-th generation. After, a combined population R_t is formed.

$$R_t = P_t \cup Q_t \tag{21}$$

R_t is sorted into different no domination levels F_j. So, we can write:

$$R_t = \bigcup_{j=1}^{r} F_j \tag{22}$$

To offer a higher precision with reduced CPU time, this proposed algorithm has been implemented using real-coded genetic algorithm [7]. So, a chromosome X corresponding to a decision variable is represented as a string of real values x_i, i.e. $X = x_1 x_2 \dots x_{lch}$ where, l_{ch} is the chromosome size and x_i is a real number within its lower limit a_i and upper limit b_i i.e. $x_i \in [a_i, b_i]$. A non-uniform arithmetic crossover is used. Thus, for two individuals having as chromosomes respectively X and Y and after generating a random number $\alpha \in [0, 1]$, the crossover operator can provide two chromosomes X' and Y' with a probability P_C as follows:

$$\begin{cases} X' = \alpha X + (1 - \alpha)Y \\ Y' = (1 - \alpha)X + \alpha Y \end{cases} \tag{23}$$

Moreover, the non-uniform mutation operator has been employed. So, at the t-th generation, a parameter x_i of the chromosome X will be transformed to other parameter x_i' with a probability P_m as follows:

$$x_i' = \begin{cases} x_i + \Delta(t, b_i - x_i) & \text{if } \tau = 0 \\ x_i + \Delta(t, x_i - a_i) & \text{if } \tau = 1 \end{cases} \tag{24}$$

with:

$$\Delta(t, y) = y \left[1 - r \left(1 - \frac{t}{g_{max}} \right)^\beta \right] \tag{25}$$

and where τ a binary is number, r is a random number and g_{max} is the maximum number of generations.

5 ANFIS approach

ANFIS was originally suggested in [18], where the ANFIS architecture was presented to model nonlinear functions, identify nonlinear components control system and also, to predict a chaotic time series.

The ANFIS is composed of five layers. A description of each layer is presented in [18]. The Takagi-Sugeno (TS) fuzzy rules are linear combinations of linear inputs [10–13].

Selection of initial number of membership functions is an important step in the ANFIS application [12]. In [19], the authors have determined this number by trial and error. They demonstrated that this method was not effective because it is based on a grid partition and it causes an explosion of the number of rules when the inputs number is large. So, they have proposed other method based on a clustering algorithm. The objective of clustering is to generate a concise representation of a system's behavior by dividing the data space into clusters. Several clustering methods are used in literature [20]. In this paper, a procedure based on subtractive clustering algorithm [20] is used to generate the initial fuzzy inference system (FIS) structure.

6 Simulation results

To evaluate the effectiveness and robustness of the proposed controller, their performances have been examined on a SMIB system given by Fig. 1. A 6-cycle fault at the infinite bus at the end of one transmission line was been considered. This fault has been cleared without line tripping.

The system data is given in the appendix. The design process of this controller has been done on two steps, which are data preparation and training phases. All time simulations are carried out using nonlinear model. In this study, lower and upper limits of the controller gains KPSS are, 10 and 100. However, minimum and maximum values of T_{1P}, T_{3P}, are respectively 0.05 and 1.

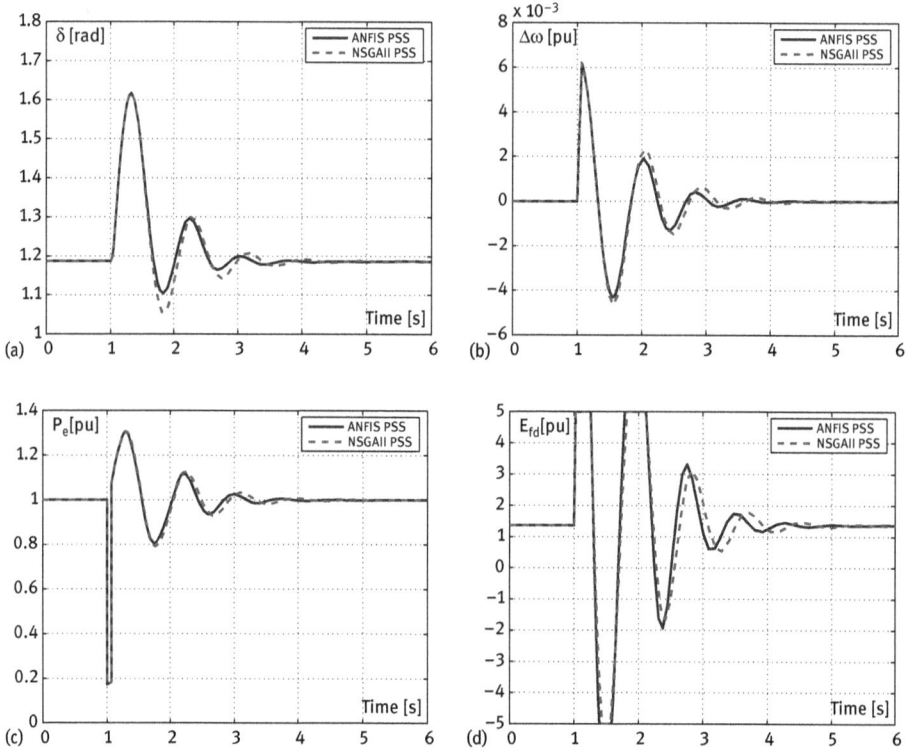

Fig. 3. (a): Rotor angle response at nominal loading. (b): Speed deviation response at nominal loading. (c): Electrical power response at nominal loading. (d): Field voltage variation at nominal loading.

6.1 Data preparation

Firstly, the NSGAII algorithm is used to tune the PSS on a wide range of operating conditions, according to the objective functions defined by equations (15) and (16). The generator real and reactive power outputs are ranged respectively, from 0.3 to 1.1 and 0.01 to 0.4 p.u. The collected input-output data pairs are stocked in a training set. Inputs are defined by the operating conditions and outputs are the corresponding PSS parameters. In this application, the training set size is 1044 input-output data pairs.

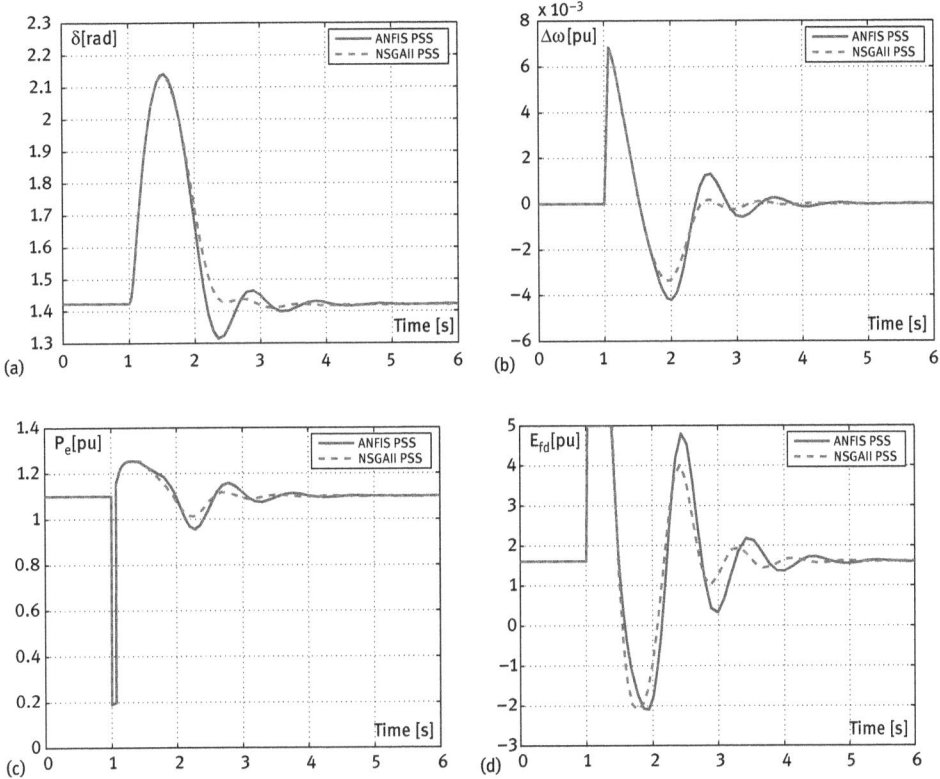

Fig. 4. (a): Rotor angle response at heavy loading. (b): Speed deviation response at heavy loading. (c): Electrical power response at heavy loading. (d): Field voltage variation at heavy loading.

In order to evaluate the effectiveness of the NSGAII algorithm, the optimal parameters of the proposed controller corresponding to a three loading conditions given in Tab. 1, are shown in Tab. 2.

Tab. 1. Loading conditions.

Loading condition	P [pu]	Q [pu]	δ_0 [rad.]
Nominal loading	1	0.015	1.1871
Light loading	0.7	0.015	0.8240
Heavy loading	1.1	0.27	1.4234

Figures 3–5 depict the nonlinear time domain simulations at the loading conditions with a six-cycle fault disturbance. It can be clearly seen that the proposed stabilizer design able to provide a good damping characteristics.

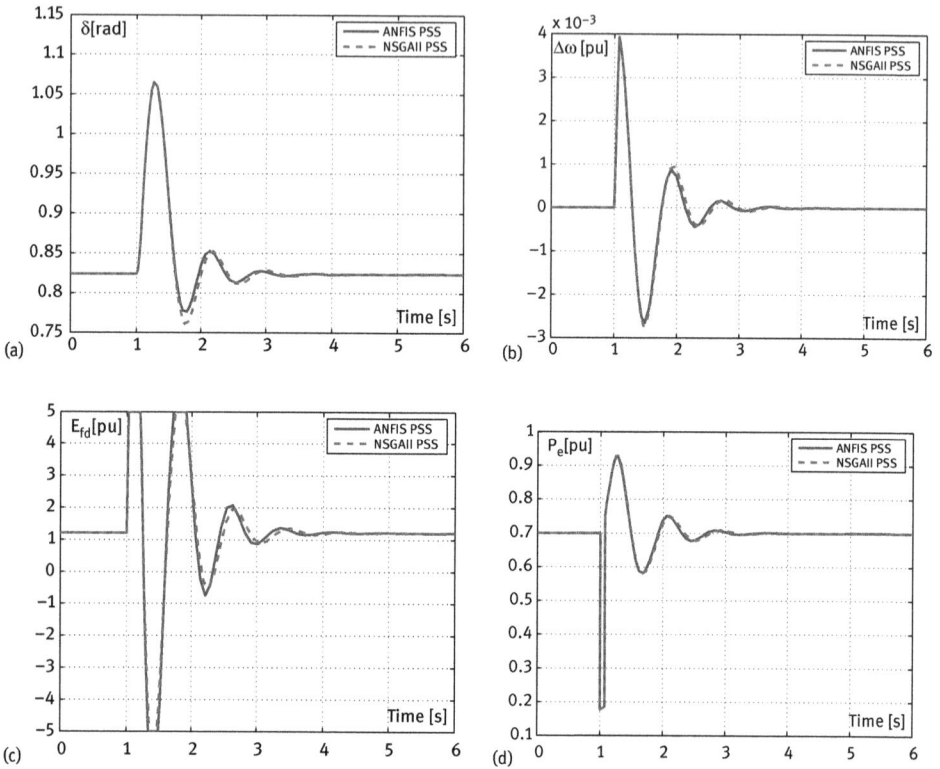

Fig. 5. (a): Rotor angle response at light loading. (b): Speed deviation response at light loading. (c): Electrical power response at light loading. (d): Field voltage variation at light loading.

Tab. 2. Optimal parameters settings of the proposed stabilizer for various loading conditions.

	K_{PSS}	T_{1P}	T_{2P}	T_{3P}	T_{4P}
Nominal loading	32.1539	0.1373	0.1000	0.1253	0.1000
Light loading	41.8617	0.1371	0.1000	0.1330	0.1000
Heavy loading	39.5598	0.0952	0.1000	0.1633	0.1000

6.2 Training phase

The initial (FIS) is trained using ANFIS, to reach the least possible error between the desired output and the FIS output through the training set. A combination of least-squares and back-propagation gradient descent methods are used.

Figure 6 shows real checking data and training data corresponding to $K_P SS$. It can be clearly seen that the ANFIS output has a good approximation.

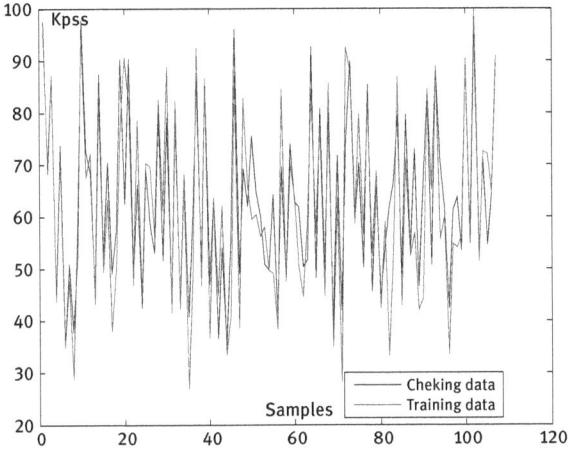

Fig. 6. ANFIS prediction and checking data.

6.3 Validation of the proposed stabilizer

To demonstrate the effectiveness of the proposed approach, closed-loop system eigenvalues at three loading conditions are computed.

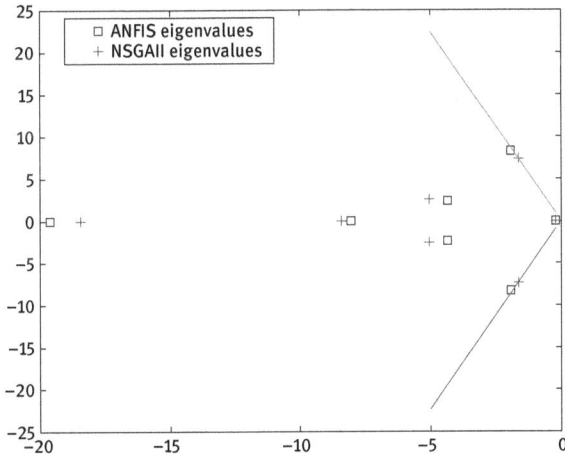

Fig. 7. D-shape sector at nominal condition.

Table 3 gives the electromechanical modes eigenvalues and their damping ratios without and with the proposed controller, at these operating conditions. It is clear that the open loop system is unstable because of the negative damping ratios. Figures 7–9 shows that all eigenvalues of the proposed stabilizer are in the D-shape sector and they are close to those obtained by the NSGAII design.

Tab. 3. System electromechanical modes without and with control. (a): Nominal loading. (b): Light loading. (c): Heavy loading.

	·No control	NSGAII optimized stabilizer parameters
(a)	$0.2949 \pm j\ 4.9576, (-0.0594)$	$-5.0454 \pm j\ 2.5537, (0.8922)$
(b)	$0.1026 \pm j\ 5.3755, (-0.0191)$	$-4.9364 \pm j\ 2.8004, (0.8698)$
(c)	$0.4786 \pm j\ 4.3491, (-0.109)$	$-4.0462 \pm j\ 1.8165, (0.9123)$

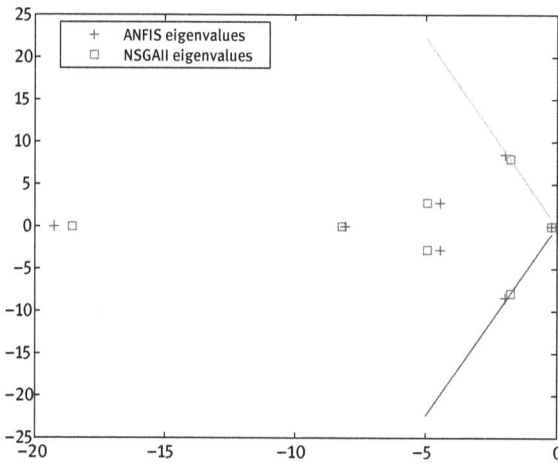

Fig. 8. D-shape sector at light condition.

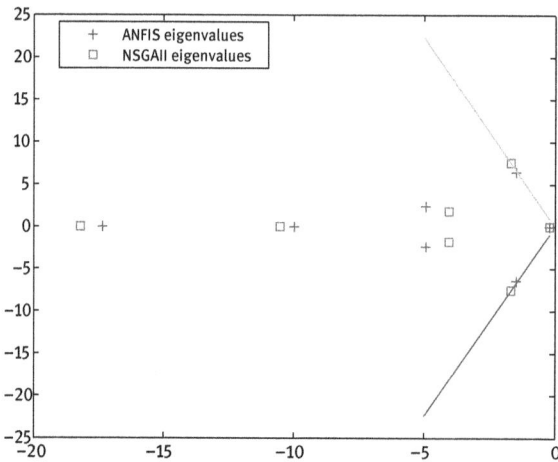

Fig. 9. D-shape sector at heavy condition.

7 Conclusion

In this study, an adaptive power system stabilizer based on adaptive network based fuzzy inference system (ANFIS) and second version of non-dominated sorting genetic algorithms (NSGA-II) is proposed. NSGA-II is used to collect the training set composed by loading conditions and corresponding stabilizer parameters. Two eigenvalue-based objective functions are considered to place the closed-loop system eigenvalues in the D-shape sector. The approach effectiveness is validated on single machine infinite bus. The nonlinear simulation results and eigenvalue analysis for various operating conditions show that the proposed stabilizer is able to provide a good damping and improves the overall system performance on real-time. The proposed approach can be applied to real time stability for multimachine power system.

Appendix: System data

Generator data:

Generator	$M = 9.26s$	$D = 0$	$x_d = 0.973$pu
	$X_q = 0.55$pu	$X'_d = 0.19$pu	$T'_{do} = 7.76s$
	$P_e = 1$pu	$Q_e = 0.015$pu	$\delta_0 = 67.61$

Exciter data:

Exciter	$K_A = 50$	$T_A = 0.05s$

Line data:

Transmission line	$R = -0.034$pu	$X = 0.997$pu
	$G = 0.249$pu	$B = 0.262$pu

Bibliography

[1]   P. Kundur. *Power System Stability and Control*. McGraw-Hill, 1994.
[2]   E.V. Larsen and D.A. Swann. Applying power system stabilizer, Part I: General concept, Part II: Performance and tunning concepts, Parts III: Practical considerations. *IEEE Trans. on Power system apparatus and systems*, PAS-100(6):3017–3024, June 1981.
[3]   M. Kashki, M. A. Abido and Y. L. Abdel-Magid. Pole placement approach for robust optimum design of PSS and TCSC-based stabilizers using reinforcement learning automata. *Electrical Engineering*, 91:383–394, 2010.

[4]   A.D. Del Rosso, C.A. Canizares and V.M. Dona. A study of TCSC controller design for power system stability improvement. *IEEE Trans. Power Systems*, 18(4):1487–1496, 2003.

[5]   Y. L. Abdel-Magid and M. A. Abido. Robust Coordinated Design of Excitation and TCSC-based Stabilizers using Genetic Algorithms. *Int. J. of Electrical Power & Energy Systems*, 69(2–3):129–141, 2004.

[6]   S. Panda, N.P. Padhy and R.N. Patel. Robust Coordinated Design of PSS and TCSC using PSO Technique for Power System Stability Enhancement. *J. Electrical Systems*, 3(2):109–123, 2007.

[7]   K. Deb. A Fast and Elitist Multiobjective Genetic Algorithm: NSGA-II. *IEEE Trans. On Evolutionary Computation*, 6(2):182–197, April 2002.

[8]   N. Srinvas and K. Deb. Multi-objective function optimization using non-dominated sorting genetic algorithms. *Evolutionary Computation*, 2(3):221–248, 1994.

[9]   C. M Fonseca and P. J. Fleming. Genetic algorithms for multiobjective optimization: formulation, discussion and generalization. 5th *Int. Conf. on Genetic Algorithms*, San Mateo California, :416-423, 1993.

[10] S.M. Radaideh, I.M. Nejdawi and M.H. Mushtafa. Design of power system stabilizers using two level fuzzy and adaptive neuro-fuzzy inference systems. *Electrical Power and Energy Systems*, 35:47–56, 2012.

[11] A.M.A. Haidar, M.W. Mustafa, F.A.F. Ibrahim and I.A. Ahmed. Transient stability evaluation of electrical power system using generalized regression neural networks. *Applied Soft Computing*, 11:3558–3570, 2011.

[12] S. R. Khuntia and S. Panda. Simulation study for automatic generation control of a multi-area power system by ANFIS approach. *Applied Soft Computing*, 12:333–341, 2012.

[13] S. R. Khuntia and S. Panda. ANFIS approach for SSSC controller design for the improvement of transient stability performance. *Mathematical and Computer Modelling*, 57:289–300, 2013.

[14] S. Panda. Differential evolutionary algorithm for TCSC-based controller design. Simulation modelling practice and theory, 17:1618–1634, 2009

[15] H. Shayeghi, A. Safari and H.A. Shayanfar. PSS and TCSC damping controller coordinated design using PSO in multi-machine power system. *Energy Conversion and Management*, 51(12):2930–37, 2010.

[16] V. Mukherjee and S.P. Ghoshal. Comparison of intelligent fuzzy based AGC coordinated PID controlled and PSS controlled AVR system. *Int. J. of Electrical Power & Energy Systems*, 29(9):679–689, 2007.

[17] H. Yassami, A. Darabi and S.M.R. Rafiei. Power system stabilizer design using Strength Pareto multi-objective optimization approach. *Electric Power Systems Research*, 80:838–846, 2010.

[18] J-S.R. Jang. ANFIS: Adaptive-network-base fuzzy inference system. *IEEE Trans. On power Systems, Man & Cybernetics*, 23(3):665–685, 1993.

[19] J. Fraile-Ardanuy and P. J. Zufiria. Design and comparison of adaptive power system stabilizers based on neural fuzzy networks and genetic algorithms. *Neurocomputing*, 70:2902–2912, 2007

[20] S. Chiu. Fuzzy Model Identification Based on Cluster Estimation. *J. of Intelligent & Fuzzy Systems*, 2(3):267–278, 1994.

Biographies

Anouar Farah received the M.S. degree in Higher School of Sciences and Techniques of Tunis from Tunis I University, Tunisia, in 1996 and Master degree in 2010. He is currently a doctoral student in the National Engineering School of Sfax, Tunisia. His research interests power system stability, FACTS, optimization techniques applied to power system and intelligent control.

Tawfik Guesmi received the electrical engineering degree (1999) and the M.Sc (2002) and Ph.D. (2007) in electrical engineering from National Engineers School of Sfax, Tunisia. He is currently an associate professor in the biomedical Institute of Tunis-Tunisia. His current research interests intelligent techniques applications in electrical power systems.

Hsan Hadj Abdallah, He received his Ph.D in electrical engineering from the Higher School of Sciences and Techniques of Tunis from Tunis I University, Tunisia, in 1991. He is currently a professor in the Department of Electrical Engineering of National School of Engineering of Sfax-Tunisia. His current research interests electrical power systems (EPS), the dispatching and the stability of EPS, wind energy, and intelligent techniques applications in EPS.

Abderrazak Ouali received his Engineering Diploma in Science physics from Tunis University in 1974. The Ph.D in automatic and informatics from 7- Paris University, in 1977. He is currently a professor in the Department of Electrical Engineering of National School of Engineering of Sfax-Tunisia. His current research interests control and electric power systems.

N. Khemiri, A. Khedher and F. Mimouni

A Sliding Mode Control Approach Applied to a Photovoltaic System operated in MPPT

Abstract: In this paper, the modeling and control of a photovoltaic system is presented. A maximum power point tracking (MPPT) algorithm is adopted to maximize the photovoltaic output power. The proposed strategy is based on the sliding mode control is robust to environment changes (irradiance and temperature). Simulation results using Matlab/Simulink have shown good performances of the photovoltaic system operated in MPPT.

Keywords: Photovoltaic system, MPPT, sliding mode control, environment changes.

Mathematics Subject Classification 2010: 65C05, 62M20, 93E11, 62F15, 86A22

1 Introduction

Used to produce electrical energy, renewable energy sources such as solar photovoltaic energy is fast-growing. The photovoltaic generator is composed of a various number of solar cells connected like series and parallels. However, the performance of photovoltaic depends on irradiation, temperature and load impedance [1]. Several literatures deal with the problem concerning the search of optimal operating point by using some maximum power point tracking (MPPT) methods in order to extract the maximum energy from the PV modules [2–4].

Therefore, the basic structure of photovoltaic system presented can contain the following components: solar PV array, with a number of series/parallel interconnected solar modules, a DC/DC boost converter, a load and a control system, illustrated by Fig. 1.

The DC-DC boost converters are extensively used in photovoltaic generating systems as an interface between the photovoltaic array and the load, which allows tracking of maximum power point. There are several methods of PV conversion with maximum power point tracking (MPPT) [5, 6]. Some of the methods use perturbation and observation method [7], incremental conductance method [8], constant voltage method [9], backstepping technique and sliding mode control (SMC) [10].

N. Khemiri, F. Mimouni: University of Monastir, National Engineering School of Monastir (ENIM) Research unit ESIER, Monastir, Tunisia, e-mails: khemirin@yahoo.fr, Mfmimouni@enim.rnu.tn.
A. Khedher: University of Sousse, National Engineering School of Sousse (ENISo), Sousse, Tunisia.
A. Khedher: University of Sfax, National Engineering School of Sfax (ENIS), Research unit RELEV, Sfax, Tunisia, e-mails: Adel_kheder@yahoo.fr.

De Gruyter Oldenbourg, ASSD – Advances in Systems, Signals and Devices, Volume 3, 2017, pp. 55–66.
DOI 10.1515/9783110448412-004

Sliding mode control offers a very good way to implement a control action which exploits the inherent variable structure nature of DC-DC converters [11]. Sliding mode control is one of the effective nonlinear robust control approaches. This mode occurs on switching surface, and the system remains insensitive to parameter variations and disturbance.

This paper is organized as follows. In the second section, we have modeled the photovoltaic system. In the third section, we have presented and modeled the DC-DC boost converter. The fourth section is devoted to the sliding mode control for boost converter. In the last section, some simulation results are shown interesting obtained control performances of the PV in terms of the robustness against environment changes.

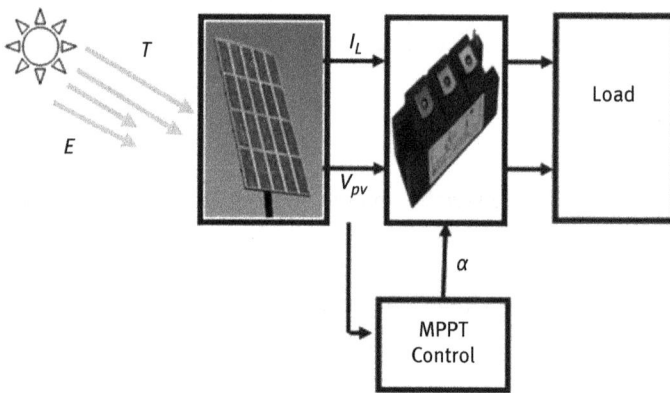

Fig. 1. Basic structure of photovoltaic system.

2 Modeling of photovoltaic systems

The electrical equivalent circuit of a solar cell is given by Fig. 2.

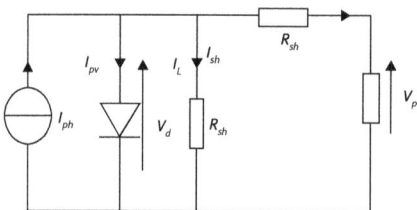

Fig. 2. Equivalent circuit model of PV.

The voltage current characteristic equation of a solar cell is given by [12]:

$$I_{pv} = I_{ph} - I_s \left[\exp\left(\frac{V_{pv} + R_s I_{pv}}{V_T}\right) - 1 \right] - \frac{V_{pv} + R_s I_{pv}}{R_{sh}} \quad (1)$$

where R_s is relatively small and R_{sh} is relatively large, which are neglected in the equation in order to simplify the simulation. The obtained ideal equivalent circuit of the PV cell is given by Fig. 3 [13].

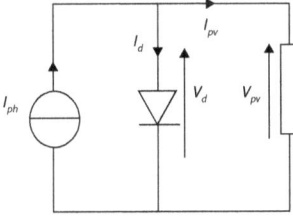

Fig. 3. Simplified electric model of a photovoltaic.

We have rewritten equation (1) as follows:

$$I_{pv} = I_{ph} - I_s \left[\exp\left(\frac{V_{pv}}{V_T}\right) - 1 \right] \quad (2)$$

In this section, we study the influence of irradiance and temperature on the photovoltaic generator. The current-voltage $I_{pv}(V_{pv})$, the power-voltage $P_{pv}(V_{pv})$ curves under different irradiance and temperature, are illustrated by Fig. 4 and Fig. 5, respectively. These figures show that the open circuit voltage is dominated by the temperature and irradiance has an influence on the short circuit current. We conclude that high temperatures and low irradiance will reduce the capacity of power conversion.

3 Mathematical model of boost DC-DC converter

Figure 6 shows the configuration of a boost DC-DC converter used to interface the photovoltaic generator with the load.

The state-space averaged model of the boost converter is given by [13]:

$$\begin{cases} \dot{x}_1 = -a_2(1-\alpha)x_2 + a_2 V_{pv} \\ \dot{x}_2 = -bx_2 + a_1(1-\alpha)x_1 \end{cases} \quad (3)$$

where: $x_1 = I_L$, $x_2 = V_s$, $a_1 = \frac{1}{C}$, $a_2 = \frac{1}{L_{pv}}$, $b = \frac{1}{RC}$ and α is the duty cycle.

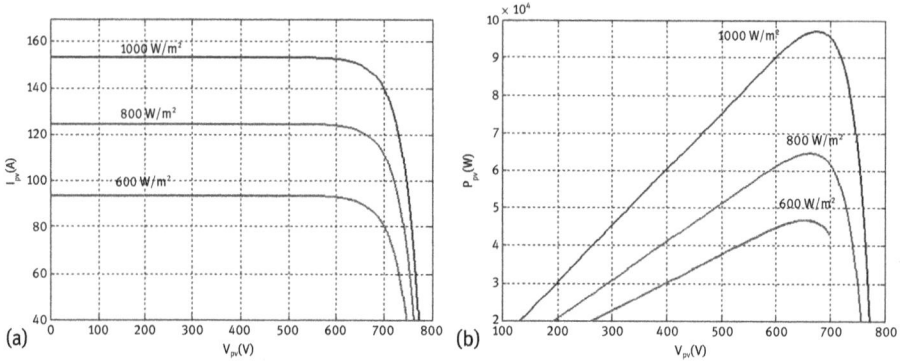

Fig. 4. (a): Model $I_{pv}(V_{pv})$, (b): Model $P_{pv}(V_{pv})$ curves for various irradiation levels: ($E = 1000, 800$ and 600 W/m²; and $T = 298°$K).

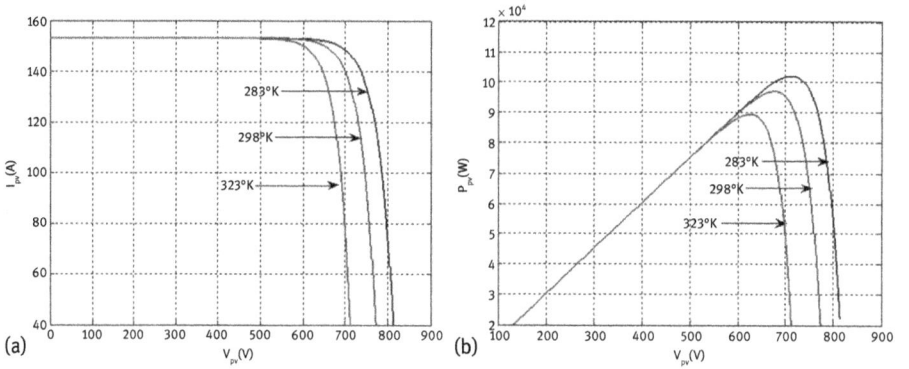

Fig. 5. (a): Model $I_{pv}(V_{pv})$, (b): Model $P_{pv}(V_{pv})$ curves for various irradiation levels: ($E = 1000$W/m², and $T = 283°$K, $298°$K and $323°$K).

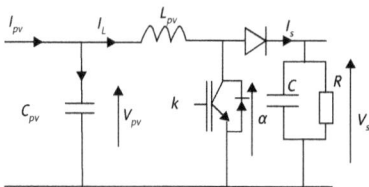

Fig. 6. Boost Converter Circuit.

The general form of the nonlinear time invariant system is given by:

$$\dot{x} = f(x) + g(x)\alpha \tag{4}$$

4 Sliding mode control of boost converter

The sliding mode control is insensitive to externals changes. In this study, we introduce the concept of the control. The sliding surface is chosen as follows [14]:

$$s = \frac{\partial P_{pv}}{\partial I_{pv}} \tag{5}$$

with: $P_{pv} = R_{pv}I_{pv}^2$. The sliding mode is obtained by the following condition [14]:

$$s = I_{pv} \left(2R_{pv} - I_{pv}\frac{\partial R_{pv}}{\partial I_{pv}} \right) = 0 \tag{6}$$

In this condition, the system operates at maximum power point of the PV generator. Based on the observation of the sliding surface with respect to the duty cycle and power photovoltaic, given by Fig. 7, we find that the sliding surface is negative for a big duty cycle and is positive for a small duty cycle.

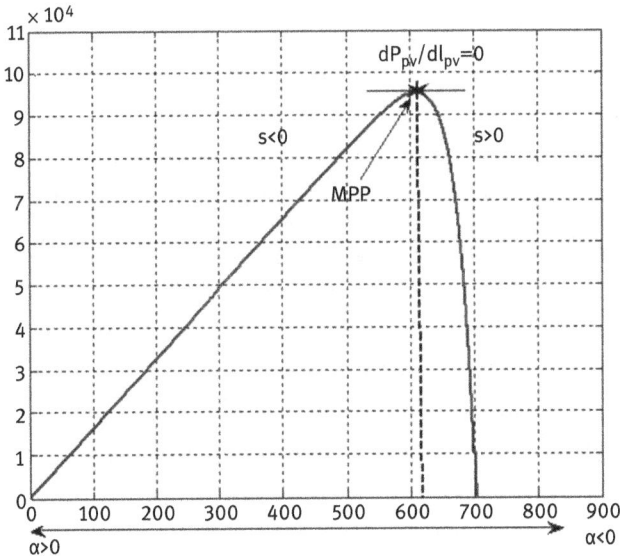

Fig. 7. Evolution of the sliding surface with respect to the duty cycle.

The evolution of the duty cycle α can be selected as the following:

$$\alpha(t+\Delta t) = \begin{vmatrix} \alpha(t)+\Delta\alpha & \text{if } s > 0 \\ \alpha(t)-\Delta\alpha & \text{if } s < 0 \end{vmatrix} \tag{7}$$

So, we can deduce the expression of the control duty cycle α which is composed of two terms α_{eq} and α_n, of which who guarantee and depended in sign of can be regarded as an effort to reach the maximum power:

$$\alpha = \alpha_{eq} + \alpha_n \tag{8}$$

The calculation of requires the application of the approach studied by [15], the equivalent control is given according to the condition defining the sliding mode.

$$\dot{s} = \left[\frac{\partial s}{\partial x}\right]^T \dot{x} \tag{9}$$

The derivative of the sliding surface can be written:

$$\dot{s} = \left[\frac{\partial s}{\partial x}\right]^T [f(x) + g(x)\alpha_{eq}] \tag{10}$$

Then, we deduce:

$$\alpha_{eq} = -\frac{\left[\frac{\partial s}{\partial x}\right]^T f(x)}{\left[\frac{\partial s}{\partial x}\right]^T g(x)} = 1 - \frac{V_{eq}}{V_s} \tag{11}$$

We choose the nonlinear law form by using the method of study by Gao [16]:

$$\alpha_n = -k|s|^n \text{sign}(s) \tag{12}$$

where k is a positive constant scalar and $0 < n < 1$. The control α is:

$$\alpha = \begin{vmatrix} 1 & \text{if} & \alpha_{eq} + \alpha_n > 1 \\ \alpha_{eq} + \alpha_n & \text{if} & 0 \le \alpha_{eq} + \alpha_n \le 1 \\ 0 & \text{if} & \alpha_{eq} + \alpha_n < 0 \end{vmatrix} \tag{13}$$

The SMC approach is based on the Lyapunov stability theory:

$$s\dot{s} \le 0 \tag{14}$$

The derivative of the sliding surface is given by:

$$\dot{s} = \left(3\frac{\partial R_{pv}}{\partial I_{pv}} + I_{pv}\frac{\partial^2 R_{pv}}{\partial I_{pv}^2}\right)\left(-a_2(1-\alpha)x_2 + a_2 V_{pv}\right) \tag{15}$$

From $R_{pv} = \dfrac{V_{pv}}{I_{pv}}$, we can write:

$$\begin{cases} \dfrac{\partial R_{pv}}{\partial I_{pv}} &= \dfrac{\partial}{\partial I_{pv}}\left[\dfrac{V_{pv}}{I_{pv}}\right] &= \dfrac{1}{I_{pv}} \times \dfrac{\partial V_{pv}}{\partial I_{pv}} - \dfrac{V_{pv}}{I_{pv}^2} \\[3mm] \dfrac{\partial^2 R_{pv}}{\partial I_{pv}^2} &= \dfrac{\partial}{\partial I_{pv}}\left[\dfrac{\partial R_{pv}}{\partial I_{pv}}\right] &= \dfrac{1}{I_{pv}} \times \dfrac{\partial^2 V_{pv}}{\partial I_{pv}^2} - \dfrac{2}{I_{pv}^2} \times \dfrac{\partial V_{pv}}{\partial I_{pv}} + 2\dfrac{V_{pv}}{I_{pv}^3} \end{cases} \qquad (16)$$

By (2), the PV voltage can be rewritten as follows:

$$V_{pv} = V_T \; \ln\left(\dfrac{I_{ph} + I_s - I_{pv}}{I_s}\right) \qquad (17)$$

and:

$$\begin{cases} \dfrac{\partial V_{pv}}{\partial I_{pv}} &= -V_T \dfrac{I_0}{I_{ph} + I_0 - I_{pv}} < 0 \\[3mm] \dfrac{\partial^2 V_{pv}}{\partial I_{pv}^2} &= -V_T \dfrac{I_0}{(I_{ph} + I_0 - I_{pv})^2} < 0 \end{cases} \qquad (18)$$

Substitute (16) into (15) yield

$$\left[\dfrac{\partial s}{\partial x}\right]^T = 3\dfrac{\partial R_{pv}}{\partial I_{pv}} + I_{pv}\dfrac{\partial^2 R_{pv}}{\partial I_{pv}^2} = \dfrac{1}{I_{pv}} \times \dfrac{\partial V_{pv}}{\partial I_{pv}} - \dfrac{V_{pv}}{I_{pv}^2} + \dfrac{\partial^2 V_{pv}}{\partial I_{pv}^2} < 0 \qquad (19)$$

The discussion of the different cases of duty cycle α is summarized as follows:
- For $0 < \alpha < 1$:

$$\dot{x} = -a_2(1-\alpha)x_2 + a_2 V_{pv} \qquad (20)$$

- For $\alpha = 0$:

$$\dot{x} = -a_2 V_s + a_2 V_{pv} < 0 \qquad (21)$$

  - $\alpha_{eq} = 0 \Longrightarrow V_{pv} = V_s$ which corresponds to the situation that the PV module is directly connected to the load.
  - $\alpha_{eq} > 0$ and $\alpha_{eq} + \alpha_n \le 0$: In this case: $s < 0$ and then $s\dot{s} < 0$
- For $\alpha = 1$:

$$\dot{x} = a_2 V_{pv} \qquad (22)$$

  - $\alpha_{eq} = 1 \Longrightarrow V_{pv} = 0$ and $s < 0$. Therefore, $\alpha_{eq} + \alpha_n \le 1$ which corresponds to a contradiction with the assumption $\alpha = 1$.
  - $\alpha_{eq} < 1$ and $\alpha_{eq} + \alpha_n \ge 1$: In this case: $s > 0$ and then $s\dot{s} < 0$

5 Simulations results

Simulations results are made by using the real parameters of photovoltaic power systems at 96KW and 680V. The specification of the system is given in the Appendix.

Sliding mode controller parameters are: $k = 0.3$ and $n = 0.2$. To study the comportment of the photovoltaic power system against the environment changes, we have illustrated several scenarios of simulations related to the irradiance and the temperature changes. In Fig. 8, we illustrate the environment changes (irradiance and temperature).

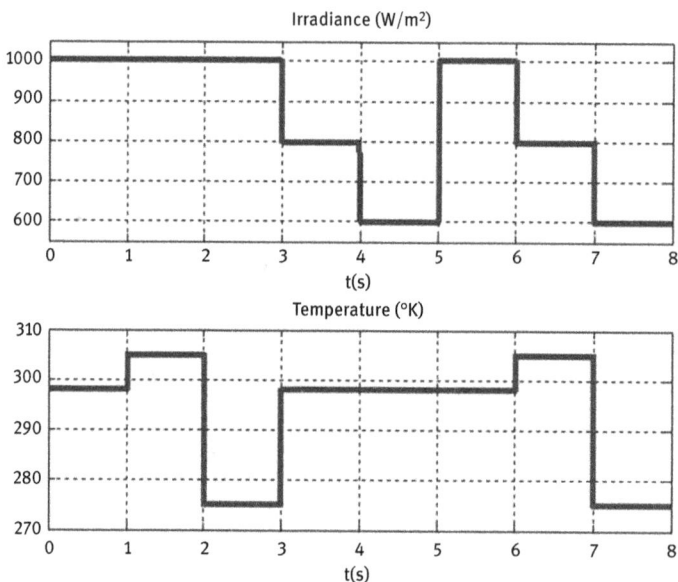

Fig. 8. Irradiance and Temperature scenario changes.

We illustrate, in Fig. 9, the inductor current I_L, the PV voltage V_{pv}, the output voltage V_s, the PV power P_{pv}, and the duty cycle α; and we illustrate in Fig. 10 the dynamic variations of the MPPT. We use the sliding mode control approach for the MPPT.

The simulation is performed by varying the temperature and the irradiance and setting the load resistance (R=30Ω).

Figures 9a and 9b are respectively the inductor current I_L (equals the current delivered by the photovoltaic I_{pv}) and the photovoltaic output voltage. The current decreases linearly with decreasing irradiation. Increasing the temperature to a value of irradiation constant time interval between 1s and 2s, causes a small decrease in PV voltage, while the current is almost constant. The consequence is manifested by a decrease in PV power (Fig. 9d).

Figure 9e shows the decrease of the irradiance amplifies the duty cycle and their increases down the duty cycle and it appears a small variation in steady condition with respect to nominal case. Thus, the simulation results confirm the robustness of the proposed scheme against the environment changes.

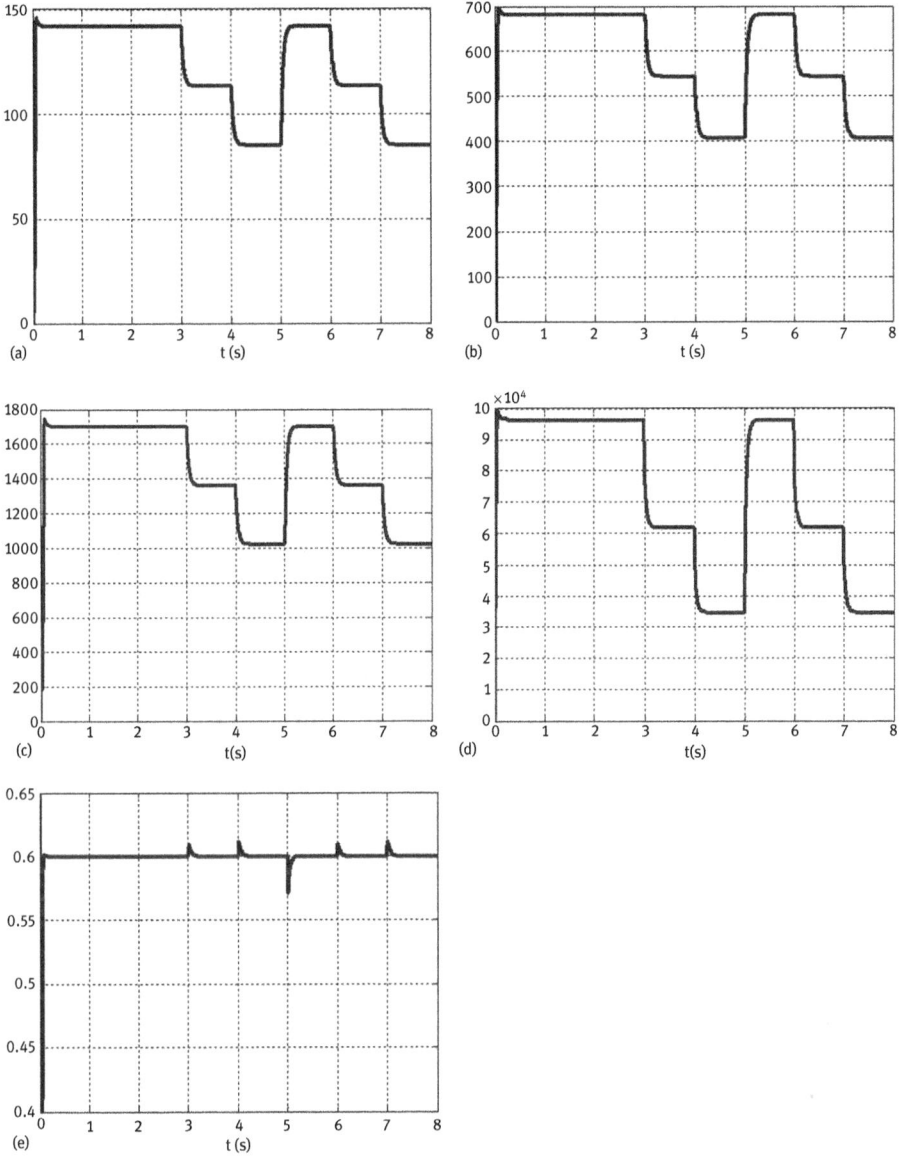

Fig. 9. PV power system operated under sliding mode strategy with irradiance and temperature variations. Legend. (a): Inductor current I_L, (b): PV voltage V_{pv}, (c): Output voltage V_s, (d): PV power P_{pv}, and (e): Duty cycle α.

Figure 10 represents the PV subsystem characteristics on the $P_{pv}(V_{pv})$ diagram. It shows the performance of the PV generation subsystem in the tracking the maximum power point MPPT of the PV during a transient of solar irradiance. The system reaches the MPP in A. Then, as a consequence of a decreasing in the solar irradiance, the operating point moves to the new MPP in B and C. When the irradiance increase, the operating point moves to the MPP in D and A. Once the new MPP is reached, only very small oscillations persist around the MPP.

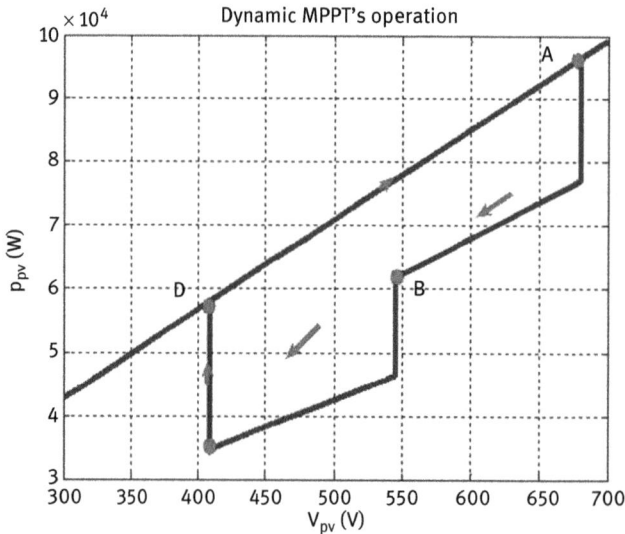

Fig. 10. Dynamic MPPT's operation.

6 Conclusion

In this paper, the photovoltaic system and the DC-DC boost converter have been described and modeled. A new MPPT method based on the sliding mode control was developed. In this context, we vary the temperature and the irradiance together, as we fix the load resistance with a value equal to 30Ω. The simulation results show on the one hand, the robustness of SMC against environment changes and on the other hand, the good performances of the photovoltaic system operated in MPPT.

Appendix

The system specification is given by the following parameters:

$$N_S = 59 \qquad N_p = 44 \qquad C_{pv} = 0.01F \quad L_{pv} = 0.03H$$
$$q = 1.6 \times 10^{-19} \quad k = 1.38 \times 10^{-23} \quad C = 68\mu F \qquad R = 30\Omega$$

Bibliography

[1]   Y. Sami, J. Moncef and H. Najib. Modelling and control of hybrid renewable energy system connected to AC grid. *Int. J. on Computer Science and Engineering* (IJCSE), 3(12), December 2011.

[2]   R.A. Mastromauro. Control Issues in Single-Stage Photovoltaic Systems: MPPT, Current and Voltage Control. *IEEE Trans. on Industrial Informatics*, 8(2):241–254, May 2012.

[3]   E. Bianconi. A Fast Current-Based MPPT Technique Employing Sliding Mode Control. *IEEE Trans. on Industrial Electronics*, 60(3):1168–1178, March 2012.

[4]   A. Khedher, M.F. Mimouni, A. Masmoudi and N. Derbel. A survey on modeling, estimation and on-line adaptation of induction motor parameters under R.F.O.C. *Trans. on Systems, Signals and Devices, Issues on Power Electrical Systems*, TSSD, 2(2):177–195, 2007.

[5]   S.L. Brunton, C.W. Rowley, S.R. Kulkarni. Maximum Power Point Tracking for Photovoltaic Optimization Using Ripple-Based Extremum Seeking Control. *IEEE Trans. on Power Electronics*, 25(10):2531–2540, 2010.

[6]   D. Casadei, G. Grandi, and C. Rossi. Single-phase single-stage photovoltaic generation system based on a ripple correlation control maximum power point tracking. *IEEE Trans. on Energy Conversion*, 21(2):562–568, 2006.

[7]   M. Azah and S. Hussain. Hopfield Neural Network Optimized Fuzzy Logic Controller for Maximum Power Point Tracking in a Photovoltaic System. *Int. J. of Photoenergy*, Article ID 798361, 2012.

[8]   W.F. Mohammad, K.M. Al-Aubidy and M.M. Ali. Maximum power point neuro-fuzzy tracker for photovoltaic arrays. *Trans. on Systems, Signals and Devices, Issues on Power Electrical Systems*, TSSD, 4(2):1–6, 2011.

[9]   A. Khedher and M.F. Mimouni. Robust Sensorless-Adaptive Nonlinear Control of Double Star Induction Machine. *Trans. on Systems, Signals and Devices, Issues on Power Electrical Systems*, TSSD, 6(2):229–245, 2011.

[10] G. Hanifi. Sliding mode control of DC-DC boost converter. *J. of Applied sciences*, 5(3):588–592, 2005.

[11] P. Mattavelli, L. Rossetto, G. Spiazzi and P. Tenti. Sliding mode control of SEPIC converters. *Proc. of European Space Power Conf.* (ESPC), Graz, :173–178, August 1993.

[12] M. Youjie, C. Deshu and Z. Xuesung. Hybrid Modeling and Simulation for boost converter in Photovoltaic system. *IEEE Int. Conf. on Information and Computer Science*, :85–87, January 2009.

[13] A. Al-Nabulsi. Efficiency Optimization of a DSP-Based Standalone PV System Using Fuzzy Logic and Dual-MPPT Control. *IEEE Trans. on Industrial Informatics*, 8(3):573–584, August 2012.

[14] R. Stala. Individual MPPT of photovoltaic arrays with use of single-phase three-level diode-clamped inverter. *IEEE Int. Symp. on Idustrial Electronics* (ISIE), :3456–3462, July, 2010.

[15] J.J.E. Slotine and W. Li. *Applied Nonlinear Control*. Prence Hall, USA, 1998.
[16] James C.H. Gao. Variable Structure control of Nonlinear systems: A new approach. *IEEE Trans. on Industrial Electronics*, 40(1), Februray, 1993.

Biographies

Nihel Khemiri was born in Mahdia, Tunisia in 1981. He received the Electrical Engineering. and Master degrees from the Tunis Engineering School (TES), in 2005 and 2006, respectively, the Ph.D. and the academic accreditation degree in Electrical Engineering from the Monastir Engineering School (MES) in 2013. In 2013, she has been an assistant professor and an associate professor in the Electronic Engineering Department of High Institute of Applied Sciences and Technology (HIAST).

Adel Khedher was born in Oued Béjà, Mahdia, Tunisia in 1967. He received the B.S. and Master degrees from the ENSET of Tunis, in 1991 and 1994, respectively, the Ph.D. and the academic accreditation degree in Electrical Engineering from the Sfax Engineering School (SES) in 2006 and 2012, respectively. From 1995 to 2002, he has been a training teacher in the professional training centers. From 2003 to 2010, he has been an assistant professor and an associate professor in the Electronic Engineering Department of High Institute of Applied Sciences and Technology (HIAST). Since November 2012 he has been promoted to Full professor in the Electronic Engineering Department of National Engineering School of Sousse. His main research interests cover several aspects related to the control of the static inverters, the electric machine drives and the renewable energy systems.

Mohamed Faouzi Mimouni received the Ph.D and University habilitation degrees in Electrical Engineering Department at Monastir-Tunisia in 1997 and 2004 respectively and is currently a Professor. His specific research interests are in the area Power Electronics, Motor Drives, Solar and Wind Power generation.

M. Buschendorf, J. Weber and S. Bernet

Comparison of the IGCT and IGBT for the Modular Multilevel Converter in HVDC Applications

Abstract: This paper briefly describes the modeling of the modular multilevel converter. The mathematical model is used to calculate the average losses and junction temperature of the semiconductor devices in dependence on the converter output power. The junction temperature, the semiconductor losses and their several components are shown in order to compare a converter with IGCTs to a converter with IGBTs. Simulation results showed that the two technologies have similar efficiency around 1200 MW; at lower power, IGBTs have higher efficiency, and at higher power, IGCTs have higher efficiency.

Keywords: modular multilevel converter, IGCT, IGBT, losses.

Mathematics Subject Classification 2010: 65C05, 62M20, 93E11, 62F15, 86A22

1 Introduction

In HVDC applications, two general converter topologies are known. Current source converters (CSC), which are also called line-commutated converters (LCC), use thyristor valves which only have switch-on capability. Thyristor CSCs either sink reactive power or require bulky circuits enabling current turn-off. Hence, they are not suitable for weak grids and do not have black start capability. Thyristor valves can be connected in series to reach voltages up to 800 kV, as realized in China between Xianjiaba and Shanghai as a bipolar transmission line with a rated power of 6400 MW [1–3]

The other general converter type is the voltage source converter (VSC), also known as the self-commutated converter (SCC), which has some significant advantages compared to the CSC. The main advantages (see [2–4]) are:
- black start capability,
- simultaneous, independent control of active and reactive power,
- operation is possible in weak networks, due to minimal need for short-circuit capacity,
- reduced footprint compared to LCC.

M. Buschendorf, J. Weber and S. Bernet: Power Electronics Group - Technische Universität Dresden, Helmholtzstraße 9, 01069 Dresden, Germany, email martin.buschendorf@tu-dresden.de.

De Gruyter Oldenbourg, ASSD – Advances in Systems, Signals and Devices, Volume 3, 2017, pp. 67–82.
DOI 10.1515/9783110448412-005

The main disadvantages of the VSC compared to the CSC are
- more complex mechanical design and
- stronger limitation in maximum rated valve current and therefor transferable power.

Possible VSC converter topologies for HVDC applications are two level converters with series connected power semiconductors or multilevel topologies, such as the
- diode-clamped (neutral-point-clamped) VSC,
- flying capacitor VSC,
- cascaded H-bridge converter and
- modular multilevel converter (MMC).

Comparisons between multilevel and two level topologies can be found in [5] and with more detail in [6]. The MMC is currently one of the most common multilevel converter topology for HVDC applications. It was introduced in 2003 by Rainer Marquardt [7]. A block diagram of the converter and transmission system for an offshore windfarm is shown in Fig. 2. VSCs employ semiconductor devices which are able to switch on and off. For multilevel applications such as HVDC, IGBT modules are usually used, but other options include IGCT, GCT, GTO, and press-pack IGBTs, with operating characteristics shown in Fig. 1. It can be seen that IGCTs are an interesting alternative

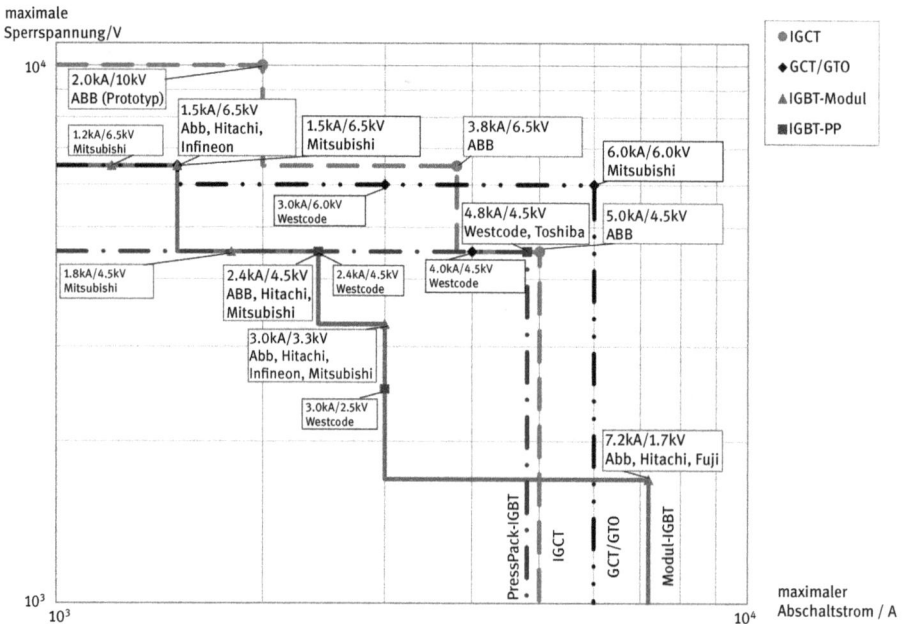

Fig. 1. Overview of power semiconductor devices with high current and voltage rating.

from the perspective of voltage blocking and current switching ratings. The blocking voltages of IGCTs up to 10 kV exceed the boundaries of IGBTs. Also the switched current rating is about twice as high, e.g. for 4.5 kV IGBT modules in comparison to IGCTs of the same nominal voltage.

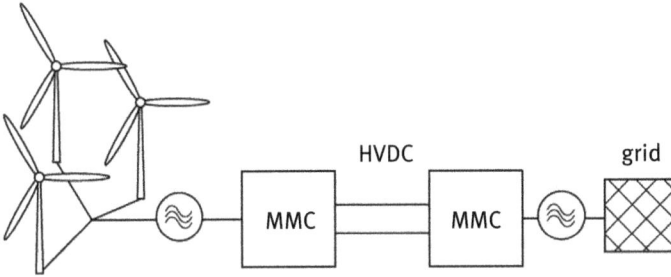

Fig. 2. Topology of a HVDC power transmission line.

Tab. 1. Characteristics of IGCTS and IGBT modules.

	IGCT	IGBT
Application	Medium voltage	Low and medium voltage
Scalability	Parallel and series (with snubber)	Parallel and series
Losses	Low	Medium
Cost	Low	Low
Type of failure	Short circuit	Open circuit (Plasma)
Chipdesign	Monolithic	Single Chips, internal parallel connection
Gate control	no active gate control	Active di/dt and du/dt control
External Clamp	Required	Possible
On-state voltage	Lower than IGBT	Higher than IGCT

Since IGBT press-pack devices are not readily available, converters with IGCT semiconductors are a common alternative to IGBT modules in industry applications with high power and current ratings. Hence this paper compares a MMC for HVDC IGBT modules to one using IGCTs in terms of losses and thermal behavior. A short overview of the characteristics of IGCTs and IGBT modules is given in Tab. 1 as can be found in [8, 9].

This paper is organized as following: Section 2 deals with the modeling of the MMC for the investigations. Section 2.1 gives the necessary equations for voltages and currents. In Section 2.2, losses and junction temperatures are calculated. Section 3 shows relevant results and Section 4 gives a conclusion. The topics of safety aspects, redundancy, and failure behavior are relevant for the practical operation of the MMC, but are not addressed by this paper.

2 Modeling

2.1 Mathematical model of the MMC converter

In Fig. 3 the equivalent circuit of the converter is shown and the definitions of used variables and parameters are given.

The converter is modeled as six controllable voltage sources, where each voltage source represents the n submodules of one phase leg. The voltages in the simulation are given as a continuous value as described in [10]. Each submodule consists of one capacitor and two semiconductor switches, each with an inverse diode. In normal operation the output voltage of a submodule can assume one of two values: $u_{SM} = u_C$, which is called "on" or activated, and $u_{SM} = 0$, which is called "off" or deactivated. The mathematical description of the converter is as follows:

$$u_{kl[z]} = \sum_{m=1}^{n} u_{SM[z,m]} = \sum_{m=1}^{n} s_{[z,m]} \cdot u_{C[z,m]}, \tag{1}$$

$$\text{for } z \in 1, 2, 3, 4, 5, 6$$

$$u_{kl[1]}(t) = \frac{1}{2} U_{dc} - U_{ac} \cos(\omega_{ac} t) + L_z \frac{di_{[1]}}{dt} \tag{2}$$

$$i_{[1]}(t) = \frac{1}{3} I_{dc} + \frac{1}{2} I_p \cos(\omega_{ac} t) + \frac{1}{2} I_q \sin(\omega_{ac} t) \tag{3}$$

where the capacitor voltage is $u_{C[z,m]}$, the corresponding switching function is $s_{[z,m]}$, the submodule voltage of the m-th submodule in phase leg z is $u_{SM[z,m]}$ the voltage of the z-th phase leg is $u_{kl[z]}$, the current in phase leg z is $i_{[z]}$, the number of activated submodules in the z-th phase leg is $m_{on[z]}$, the amplitude of the grid voltage is U_{ac}, and the radial frequency of the grid is ω_{ac}. I_p and I_q are the amplitudes of the real and the imaginary components of the grid current, respectively. The current $i_{[z]}$ is directly defined by the DC (first term on the right hand side of (3) and AC (second and third term of the right hand side of (3)) current components at the nodes of each phase leg. For this model, the converter is assumed to be lossless, and it follows that the sum of the submodule output voltages is equal to the phase leg voltage [10]. The ratio of how many of the submodules are activated is expressed by the drive factor

$$m_{in[z]} = \frac{m_{on[z]}}{n}. \tag{4}$$

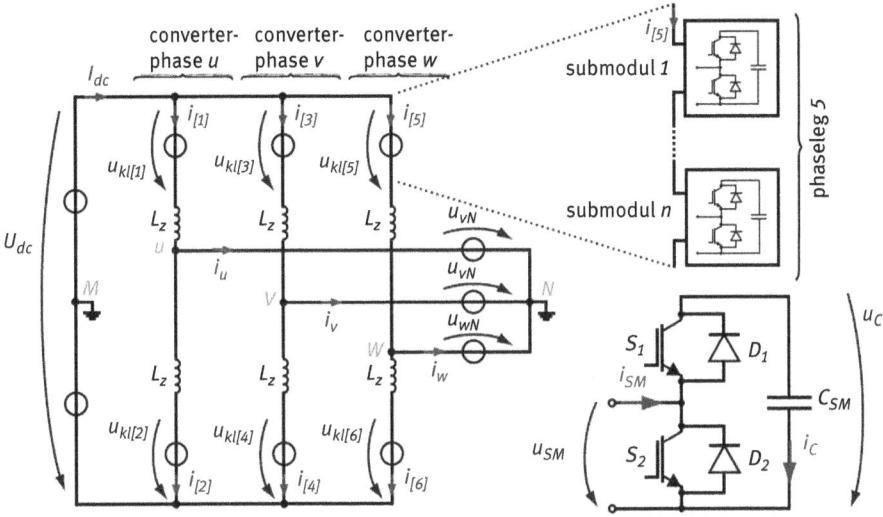

Fig. 3. Equivalent circuit of the modular multilevel converter (MMC).

2.2 Model of the semiconductors to calculate the losses and junction temperature

2.2.1 Conduction losses

The commonly used equation for the calculation of the conduction losses

$$P_{V,\text{cond}} - \frac{1}{T_1} \int_{t_0}^{t_0+T_1} u_{\text{cond}}(t) i_{\text{cond}}(t) \mathrm{d}t \tag{5}$$

is used, where u_{cond} and i_{cond} are the device terminal voltage and current during conduction. The losses of each semiconductor (as labeled in Fig. 3) in the submodule depend on phase leg current direction, and are scaled by the drive factor as given in Tab. 2.

Tab. 2. Conduction losses of semiconductor devices in the MMC submodule.

	S_1	DS_1	S_2	DS_2
$i_{[1]} > 0$	0	$(1-m_{\text{in}[1]})$ $\times P_{V,\text{cond}}$	$m_{\text{in}[1]}$ $\times P_{V,\text{cond}}$	0 0
$i_{[1]} < 0$	$m_{\text{in}[1]}$ $\times P_{V,\text{cond}}$	0 0	0	$(1-m_{\text{in}[1]})$ $\times P_{V,\text{cond}}$

For positive phase leg current, devices S_1 and D_2 are blocking and do not conduct current. For negative phase leg current, devices S_2 and D_1 are blocking. The distribution of the conduction losses amongst the semiconductor devices also depends on drive factor $m_{in[z]}$ since this determines the actual phase leg voltage in relation to the maximum phase leg voltage.

2.2.2 Switching losses

The switching losses are also computed as commonly done, by integrating the mean switching energy over one switching period, scaled by the ratio of the actual to nominal voltage, and multiplied by the switching frequency:

$$P_{V,sS} = f_{switch} \frac{1}{T_1} \int_{t_0}^{t_0+T_1} \left[E_{on}(t) + E_{off}(t) \right] \frac{u_{c(t)}}{U_{nom}} dt \tag{6}$$

$$P_{V,sD} = f_{switch} \frac{1}{T_1} \int_{t_0}^{t_0+T_1} E_{rec}(t) \frac{u_{c(t)}}{U_{nom}} dt \tag{7}$$

For an IGCT and its inverse diode, the switching energies are given in the device datasheet and scaled as described by (6) and (7). For IGBT modules, the switching energies E_{on}, E_{off}, and E_{rec} were determined by experiment. In switching tests the energies for different operating conditions were measured and a function $E = f(U, I)$ was determined which approximates the relations.

A general switching frequency cannot be given, because of the nondeterministic switch control. For this investigation, an average switching frequency of switch $f_{switch} = 150\,Hz$ is assumed.

2.2.3 Losses in clamp circuit

Additionally to the conduction and switching losses of the semiconductors, the IGCT has a relevant amount of losses in the mandatory clamp as shown in Fig. 4. For the calculation the prior losses are those caused by the energy stored in the clamp inductivity, which has to be wasted every switching period within the clamp resistor.

These losses are calculated as done in [11] per

$$P_{V,Cl} = f_{switch} \frac{1}{T_1} \int_{t_0}^{t_0+T_1} \left(\frac{L_{Cl}}{2} i_l^2(t) \right) dt. \tag{8}$$

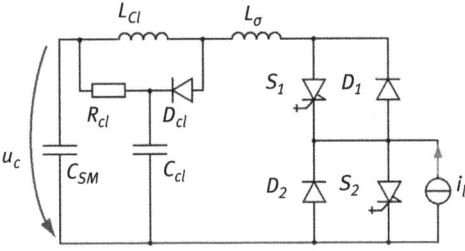

Fig. 4. Clamp circuit for IGCT configuration.

2.2.4 Thermal model

Based on the determined semiconductor losses a thermal model is applied to calculate the junction temperatures of the semiconductor devices. It is adapted as described in [12] for thermal equivalent circuits and follows the physical cauer model, and is simplified by neglecting the thermal capacities and only using the thermal resistors. Hence, the model enables the calculation of the average junction temperatures. The model is not applicable for transient processes as instantaneous power variations. Thermal circuits for the IGCT and the IGBT including their inverse diodes are shown in Fig. 5.

In the thermal circuits, the resistors represent the thermal interfaces between: junction and housing (SG), housing and heat-sink (GK), and heat-sink and environment (KU), from left to right. In the IGCT converter, the diode and switch are separately installed and the thermal resistors between housing and cooling are consequently also distinct. Because of the module housing, IGBT chips and diode chips are stronger thermally coupled and therefore the thermal resistors between housing and cooling are coupled and drawn as one single element. The current sources represent device power losses and the voltage source represent the temperature difference to a reference temperature. The junction temperature is the voltage drop across the current source.

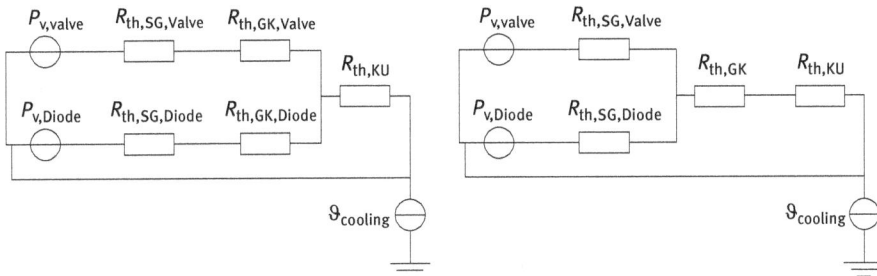

Fig. 5. Stationary thermal models for a) IGCT Press-Pack and b) IGBT module.

2.2.5 Simulation parameters

For this investigation, an IGCT for medium switching frequencies and a suitable inverse diode with 4.5 kV nominal voltage and high current rating was chosen. For comparison, a 4.5 kV IGBT module was chosen as well. The selected devices are
- ABB Semiconductors, IGCT 5SHY 35 L 4500 [13] with,
- Infineon, inverse diode D1961SH [14] and
- Mitsubishi, IGBT CM1200HC-90R [15].

The simulation parameters are summarized in Tab. 3.

Tab. 3. Simulation parameters of the HVDC converter.

U_{dc}	f_{ac}	ω_{adc}	L_{Cl}	C_{Cl}	R_{Cl}
600 kV	50 Hz	$2\pi f_{ac}$	3 μH	15 μF	1 Ω

3 Losses and junction temperature of the HVDC-MMC

The simulation is made for one representative submodule and scaled up for the entire converter. This is justified since all submodules experience practically a similar loading. This is achieved by an algorithm which selects submodules for activation/deactivation with the goal of maintaining equal submodule capacitor voltages within a phase leg [16, 17].

3.1 Junction temperature

Figure 6 shows the average junction temperature in dependence on the converter power for the two most heavily loaded semiconductors in the submodule: diode 2 and switch 2 for all realistic drive factors, $0 < m_{in[z]} > 1$.

It can be seen that the IGCT converter can achieve a maximum power of 2000 MW, whereas the IGBT exceeds its maximum junction temperature at approximately 1700 MW. The active semiconductor chip area is similar for both, which is achieved by simulating two IGBTs in parallel, a technique commonly found in the literature, for example in [18, 19]. It is apparent that, in the chosen application, the inverse diode is the limiting device for the IGCT converter as well as for the IGBT converter. The asymmetrical junction temperature is a result of the chosen drive factor which is much smaller than one to ensure safe operation.

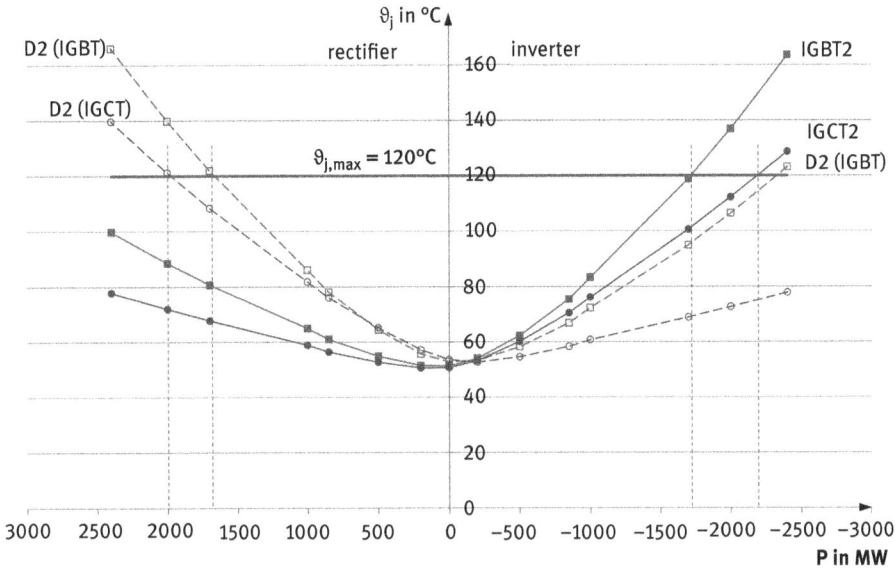

Fig. 6. Junction temperature as a function of the converter power; $Q = 350\,\text{MVar}$ for $P < 1000\,\text{MW}$, $Q = 500\,\text{MVar}$ for $P \geq 1000\,\text{MW}$, $f_{\text{switch}} = 150\,\text{Hz}$, $U_{\text{dc}} = 600\,\text{kV}$.

3.2 Semiconductor losses

The reason for the higher possible converter power for the IGCT converter is superior thermal behavior and lower semiconductor losses.

3.2.1 Losses in different operating conditions

The total semiconductor losses in dependence of the converter power are shown in Fig. 7.

In rectifier mode, semiconductor losses are higher for the IGCT converter compared to the IGBT converter. The slope of the curves are nearly the same, with only a vertical offset making IGCT losses 5.0% higher than for the IGBT converter. The reason for that is the almost same dependence between current and losses in both used inverse diodes, which outweigh the switch losses. Another interesting fact is noticeable in the inverter mode, where the slope of the IGCT curve is significantly lower. At no load, the IGCT converter still has higher losses but as the power level increases, the losses for the IGCT converter fall below those for the IGBT converter starting around 900MW. This is due of the substantially lower conduction losses of the IGCTs, and will be explained in detail in the following section.

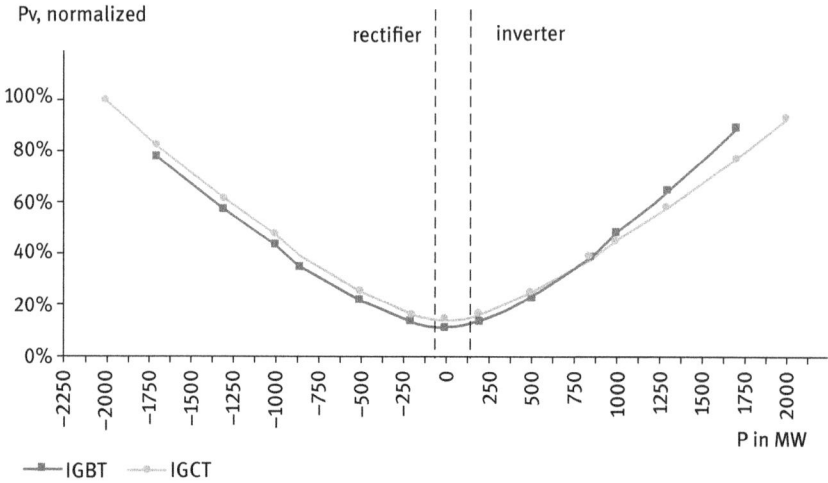

Fig. 7. Normalized semiconductor losses; $Q = 350\,\text{MVar}$ for $P < 1000\,\text{MW}$, $Q = 500\,\text{MVar}$ for $P \geq 1000\,\text{MW}$, $f_{switch} = 150\,\text{Hz}$, $U_{dc} = 600\,\text{kV}$.

3.2.2 Loss distribution

Figure 8 shows the losses of one submodule at one specific operating point, specified by transferred power $P = 850\,\text{MW}$, $Q = 350\,\text{MVar}$, switching frequency $f_{switch} = 150\,\text{Hz}$, and dc-voltage $U_{dc} = 600\,\text{kV}$, which are split into its components for the IGCT submodule as well as the IGBT configuration.

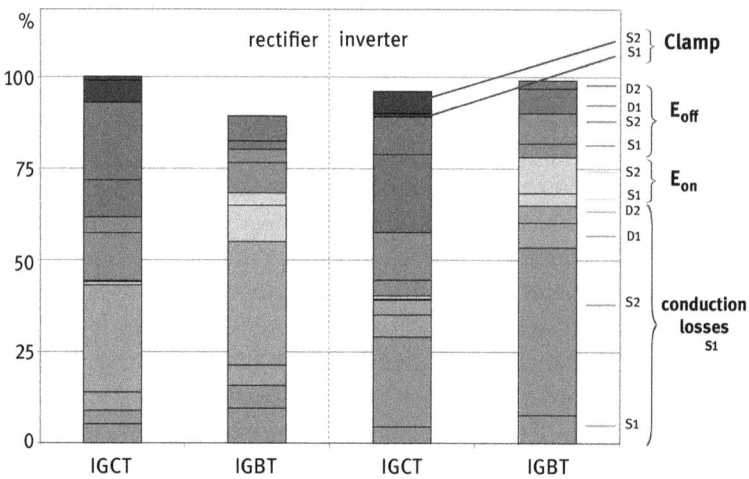

Fig. 8. Semiconductor losses of a MMC submodule divided in its components; $P = 850\,\text{MW}$, $Q = 350\,\text{MVar}$, $f_{switch} = 150\,\text{Hz}$, $U_{dc} = 600\,\text{kV}$, normalized to IGCT losses in rectifier operation.

Figures 9 and 10 show in style of Fig. 8 the losses for the operating points $P = 1700\,\text{MW}$, $Q = 500\,\text{MVar}$, $f_{\text{switch}} = 150\,\text{Hz}$, $U_{\text{dc}} = 600\,\text{kV}$ and $P = 2000\,\text{MW}$, $Q = 500\,\text{MVar}$, $f_{\text{switch}} = 150\,\text{Hz}$, $U_{\text{dc}} = 600\,\text{kV}$ where the result for the IGBT in Fig. 10 is only theoretical because of exceed junction temperature, as mentioned in Section 3.1.

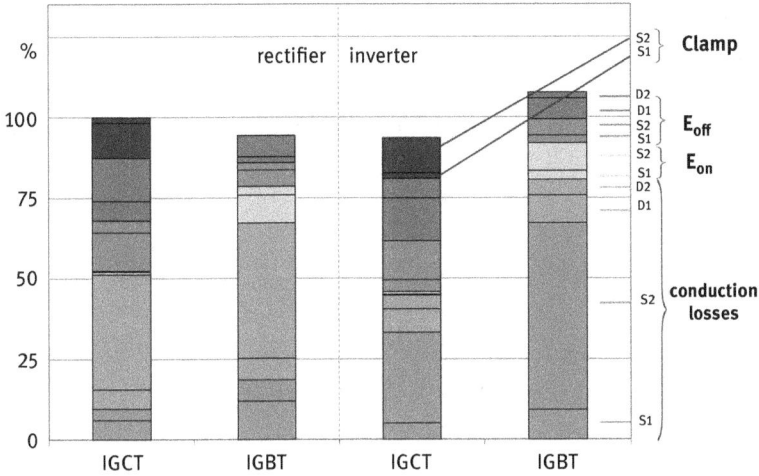

Fig. 9. Semiconductor losses of a MMC submodule divided in its components; $P = 1700\,\text{MW}$. $Q = 500\,\text{MVar}$, $f_{\text{switch}} = 150\,\text{Hz}$, $U_{\text{dc}} = 600\,\text{kV}$, normalized to IGCT losses in rectifier operation.

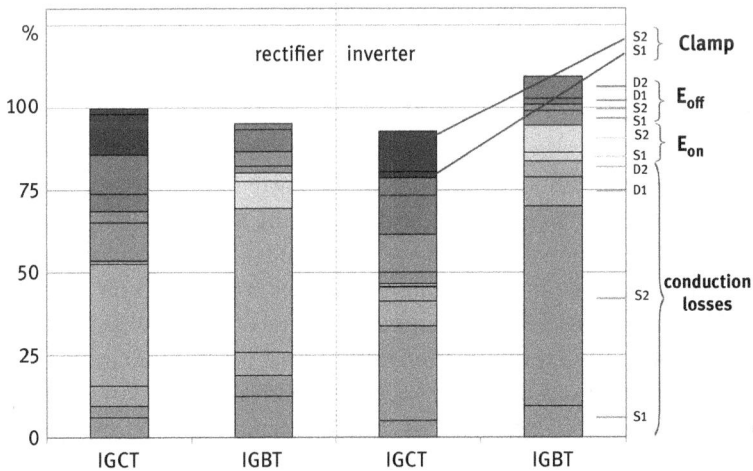

Fig. 10. Semiconductor losses of a MMC submodule divided in its components; $P = 2000\,\text{MW}$, $Q = 500\,\text{MVar}$, $f_{\text{switch}} = 150\,\text{Hz}$, $U_{\text{dc}} = 600\,\text{kV}$, normalized to IGCT losses in rectifier operation.

As expected from Fig. 6, diode 2 for rectifier mode and the IGCT 2 / IGBT 2 for inverter mode are the most stressed devices in a submodule. The second interesting effect is the distribution between conduction and switching losses. The IGCT configuration has substantially lower conduction losses compared to the IGBT configuration at the considered operating point. The switching losses for IGCT, especially for the IGCT inverse diode, are distinctly higher. Additionally the losses in the mandatory clamp circuit increase the overall losses for the IGCT converter. Finally, the overall semiconductor losses for an AC-DC-AC transmission system (including both inverter and rectifier losses) depending on converter power is shown in Fig. 11. Losses given by peripheral components, such as cooling, electronics, transmission losses, etc., are not taken into account.

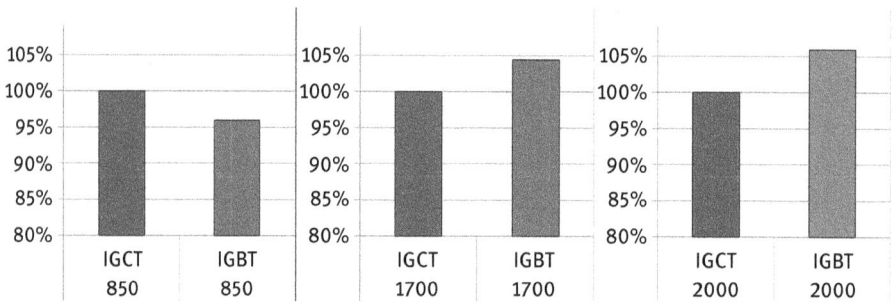

Fig. 11. Total semiconductor losses as a sum of inverter and rectifier losses; $f_{switch} = 150\,\text{Hz}$, $U_{dc} = 600\,\text{kV}$, a) $P = 850\,\text{MW}$, $Q = 350\,\text{MVar}$, b) $P = 1700\,\text{MW}$, $Q = 500\,\text{MVar}$, c) $P = 2000\,\text{MW}$, $Q = 500\,\text{MVar}$.

For transmission of 850 MW, as shown in Fig. 11 a), the IGBT implementation of the converter has nearly 4% lower semiconductor losses compared to the IGCT. This is the break-even point between IGCT and IGBT converter losses in inverter operation mode. Below this amount of converter output power the IGBT system has superior efficiency. Figure 11 b) is the point of maximum output power for the IGBT converter (1700 MW), and here the IGBT solution has nearly 4.5 % higher losses. This shows the advantages of the low conduction losses of IGCT at high currents, despite the higher losses in the inverse diode and the additionally losses in the clamp circuit. This effect increases up to 6 % by increasing the output power up to 2000 MW in Fig. 11 c) (where the IGBT already has a too high junction temperature, which could in practice be solved by using a more powerful cooling system). The break-even point between IGCT and IGBT regarding all semiconductor losses is approximately 1200 MW for the investigated devices.

4 Conclusions

In this paper, a mathematical model for the MMC was described, and for the semi-conductor losses and the junction temperature, which was then used for comparative analysis. Converter losses and junction temperatures for an MMC of a hypothetical HVDC transmission system were investigated, for which suitable high power semiconductors were selected. 4.5kV IGCTs and 4.5kV IGBT modules were chosen for the MMC submodules. The comparison of junction temperatures, semiconductor losses and loss distributions indicate that for high power, high current transmission systems the IGCT is an attractive alternative compared to IGBT modules. The IGCT converter enables up to 4% higher efficiency (for high current ratings), and higher power throughput capability for equally rated semiconductors. Especially when HVDC transmission lines are used to strengthen a heavily-loaded AC grid, as is done in San Francisco by Trans Bay Cable, the momentary power throughput is usually above 70% of capacity. In cases such as this, the use of IGCTs offers a significant advantage. However, IGBT converters have lower losses in applications which usually work at low power conditions, or with strongly fluctuating power like in connections to wind parks or photovoltaic fields.

Nomenclature

$u_{C[z,m]}$	Capacitor voltage of the m-th capacitor in the phase leg z
u_{SM}	Submodule voltage
$u_{kl[1]...[6]}$	Voltage on the submodule terminals
U_{dc}	Direct current voltage
u_{uN}, u_{vN}, u_{wN}	Phase voltages at the ac side
i_{SM}	Current trough the submodule
$i_{[1]...[6]}$	Phase leg currents
I_{dc}	Direct current
I_p	Amplitude of the real part of the phase current
I_q	Amplitude of the imaginary part of the phase current
U_{ac}	Amplitude of the alternating current voltage
ω_{ac}	Angular frequency of the alternating current
f_{ac}	Frequency of the alternating current
f_{switch}	Average switching frequency
L_z	Phase inductor
S_1	Upper IGBT in the submodule configuration
S_2	Lower IGBT in the submodule configuration
D_1	Upper diode in the submodule configuration
D_2	Lower diode in the submodule configuration
m	Momentary submodule
n	Count of submodules
z	Momentary phase leg

$m_{on[z]}$	Count of "switched on" submodules in the phase leg z
$m_{in[z]}$	Drive factor
t	Time
$P_{V,cond}$	Conduction losses
t_0	Starting time for integration
T_1	Time period of integration
u_{cond}	Semiconductor voltage during conduction
i_{cond}	Semiconductor current during conduction
$P_{V,sS}$	Switching losses of IGBTs
$P_{V,sD}$	Switching losses of diodes
E_{on}	Turn on energy
E_{off}	Turn off energy
u_c	Voltage of the submodule capacitor
U_{nom}	Nominal voltage of semiconductor
E_{rec}	Reverse recovery energy
$P_{V,Cl}$	Clamp losses
L_{Cl}	Clamp inductance
i_l	Load current
C_{Cl}	Clamp capacitor
D_{Cl}	Clamp diode
R_{Cl}	Clamp resistor
L_σ	Stray inductance
C_{SM}	Submodule capacity
P	Converter output power
Q	Converter reactive power
ϑ_j	Junction temperature
$\vartheta_{j,max}$	Maximum junction temperature
R_{th}	Thermal resistor

Bibliography

[1] A. Kumar, D. Wu, and R. Hartings. Experience from First 800 kV HVDC Test Installation *International Conference on Power Systems*, Bangalore, India, 2007.

[2] M.P. Bahrman and B.K. Johnson The ABCs of HVDC Transmission Technology. *IEEE Power & Energy Magazine*, Vol 5. No. 2, March/April 2007.

[3] J. Dorn, H. Huang, D. Retzmann. Novel voltage source converters for HVDC and FACTS applications. *Conf. CIGRE Symposium*, Osaka, CD-Rom, 2007.

[4] S. Henry, A.M. Denis, P. Panciatici. Feasibility study of off-shore HVDC grids. *IEEE Power and Energy Society General Meeting*, MN, 2010.

[5] J. Dorn, H. Huang, D. Retzmann. A new Multilevel Voltage-Sourced Converter Topology for HVDC Applications. *Cigre*, Paris, 2008.

[6] S.S. Fazel, *Investigation and Comparison of Multi-Level Converters for Medium Voltage Applications*. PhD thesis, TU Berlin, Berlin, 2007.

[7] R. Marquardt, A. Lesnicar, J. Hildinger. Modulares Stromrichterkonzept für Netzkupplungsanwendungen bei hohen Spannungen. *ETG-Fachtagung*, Bad Nauheim, 2002.

[8] P.K. Steimer, O. Apeldoorn, B. Ødegård, B. Bernet, T. Brückner. Very High Power IGCT PEBB technology. *Power Electronics Specialists Conference*, Recife, Brasil, 2005.
[9] S. Bernet, R. Teichmann, A. Zuckerberger, P.K. Steimer. Comparison of High-Power IGBT's and Hard-Driven GTO's for High-Power Inverters *IEEE Trans on industry applications*, Vol. 35, No. 2, 1999
[10] S. Rohner, J. Weber, S. Bernet. Continuous Model of Modular Multilevel Converter and Experimental Verification. *IEEE Energy Conversion Congress & Exposition (ECCE2011)*, Phoenix, Arizona, USA, September 2011.
[11] S. Tschirley. *Automatisierte messtechnische Charakterisierung von 10kV Integrierten Gate-kommutierten Thyristoren (IGCTs)*. PhD thesis, TU Berlin, Berlin, 2007.
[12] A. Wintrich, U. Nicolai, W. Tursky, T. Reimann. *Applikationshandbuch Leistungshalbleiter* ISLE Verlag, Germany, 2010.
[13] ABB Semiconductors IGCT 5SHY 35 L 4500. *datasheet*, ABB Semiconductors.
[14] Infineon Technologies AG Schnelle beschaltungslose Diode D1961SH *datasheet*, Infineon Technologies AG.
[15] Mitsubishi Electric Corporation High Voltage IGBT CM1200HC-90R *datasheet*, Mitsubishi Electric Corporation.
[16] S. Rohner, S. Bernet, M. Hiller, R. Sommer. Modulation, Losses, and Semiconductor Requirements of Modular Multilevel Converters. *IEEE Transactions on Power Electronics*, 57(8): 2633–2642, 2010.
[17] H. Akagi, M. Hagiwara, R. Maeda. Theoretical Analysis and Control of the Multilevel Cascade Converter Based on Double-Star Chopper-Cells. *The 2010 International Power Electronics Conference*, Tokyo, Japan, 2010.
[18] J. Nelson, G. Venkataramanan, B. Beihoff. Investigation of parallel operation of IGBTs. *Conference Record of the Industry Applications Conference, 37th IAS Annual Meeting*, vol. 4, pp. 2585–2591, 2002.
[19] R. Alvarez, K. Fink, S. Bernet. Simulation and experimental investigation of parallel connected IGBTs. Industrial Technology (ICIT) *IEEE International Conference on Industrial Technology*, Viña del Mar, Chile, 2010.

Biographies

Martin Buschendorf received his Dipl.-Ing degree in electrical engineering from the Technische Universität Dresden, Germany, in 2011. He is currently working on his PhD thesis in the area of characterization of power semiconductor devices for HVDC applications. His main fields of research include the construction of a test bench, the calculation of stray inductances for dc-busbars and measurements of the switching behavior of power IGBTs.

Jens Weber received the Dipl.-Ing. and Dr.-Ing. degrees in electrical engineering from the Technische Universität Dresden, Germany, in 1999 and 2007, respectively. He is currently with the Professur Leistungselektronik, Elektrotechnisches Institut, Technische Universität Dresden, where he leads an industry-funded research project dealing with power supplies for high output voltages. He is also involved in several projects funded by the German Federal Ministry of Education and Research (BMBF) or industry. His research interests include the modelling of power electronics circuits, nonlinear phenomena in switched-mode power supplies and nonlinear control of power electronic systems.

Steffen Bernet received the diploma degree from Dresden University of Technology in 1990 and the Ph.D. degree from Ilmenau University of Technology in 1995, both in electrical engineering. During 1995 and 1996, he worked as Postdoc in the ECE Department of the University of Wisconsin - Madison. In 1996, he joined ABB Corporate Research, Heidelberg (Germany) where he led the Electrical Drive Systems Group. From 1999 to 2000 he was subprogram manager responsible for the ABB research in the areas "Power Electronics Systems", "Drives" and "Electric Machines". From 2001 to 2007 he was Professor for Power Electronics at Berlin University of Technology. Since June 2007 he has been Professor at Dresden University of Technology. During the past twenty years, Dr. Bernet has conducted comprehensive research on power semiconductors, static power converters and ac motor drives.

H. Kouki, M. Ben Fredj and H. Rehaoulia

Modeling of Double Star Induction Machine Including Magnetic Saturation and Skin effect

Abstract: The aim of this paper is to present a $d-q$ accurate mathematical model of the double star induction machine (DSIM), whatever the electrical shift between the two stator stars. The proposed model takes into account the effect of magnetic saturation, the stator mutual leakage inductance between two stars and the skin effect. Obtained simulation and experimental results confirm the validity and the performances of the proposed model, especially at start-up.

Keywords: Double star induction machine, modeling, main flux saturation, stator mutual leakage inductance, skin effect.

Mathematics Subject Classification 2010: 65C05, 62M20, 93E11, 62F15, 86A22

1 Introduction

Machines with more than three phases have been applied for years in industrial applications, with their performances considered primarily in the high power field [1–3]. The main advantage of increasing stator phase number is that on the one hand it allows the reduction of the size of the components in power modulators of energy, and on the other hand it provides better tolerance as well as greater reliability [4–6]. Among multiphase machines, the double star induction machine whose angular shift between the two stars 0°, 30° or 60° is the most popular in industrial applications [7]. Such machine, in addition to the power segmentation and redundancy it carries, has the advantage of reducing torque pulsations, rotor losses and current harmonics [8, 9]. In fact, the introduction of magnetic nonlinearities in the electrical equations has always been a topical issue for polyphase machines. Indeed, taking into account the saturation is not simply dictated by the desire to improve the results, but it can sometimes be a necessity [10, 11].

The presence of the mutual leakage inductance between two stars is due to the fact that their windings share the same slots [12–14] and are therefore mutually coupled. The mutual leakage coupling depends on the winding pitch and the angle shift between the two stator winding sets. Nevertheless, there have been some studies where the stator mutual leakage coupling has been neglected [15–17].

H. Kouki, M. Ben Fredj and H. Rehaoulia: University of Tunis, Institute of Sciences and Technology of Tunis, Electrical department, Tunis, emails: hajer_kouki@yahoo.fr; mouldi.benfredj@esstt.rnu.tn; habib.rehaoulia@esstt.rnu.tn.

De Gruyter Oldenbourg, ASSD – Advances in Systems, Signals and Devices, Volume 3, 2017, pp. 83–95.
DOI 10.1515/9783110448412-006

Moreover, if the machine is fed by a voltage inverter, the rotor currents contain some specific harmonics. The frequency of these harmonic currents can have very high values. Consequently, the current distribution is not uniform in the rotor bars, which is known as the skin effect phenomenon. The skin effect increases the rotor resistance and decreases the rotor leakage inductance. We can use the Foster model to describe the dependency of rotor impedance with the effects induced by the variation of frequency. Foster models are usually referred to lumped parameter models RL circuits obtained by frequency identification measures. Parameters variations can be determined using correction factors for the leakage inductance and resistance. Nevertheless, other studies provide analytical expressions for the resistance and leakage inductance versus rotor slip, which makes it easier to take into account the skin effect [18].

Furthermore, various models have been developed to study the behavior of electrical machines. There are current models and mixed models. In the present study, we have applied the model of the current taking into account the non-linearity of the magnetic circuit and the skin effect. In this paper, the transient performance of the double star induction machine during start-up has been determined. Finally, the validation of the model has been achieved by an experimental application.

2 Test and design machines

Experimental results are obtained by means of the test bench of Fig. 1, where necessary requirements for machine supply and signals measurements are provided.

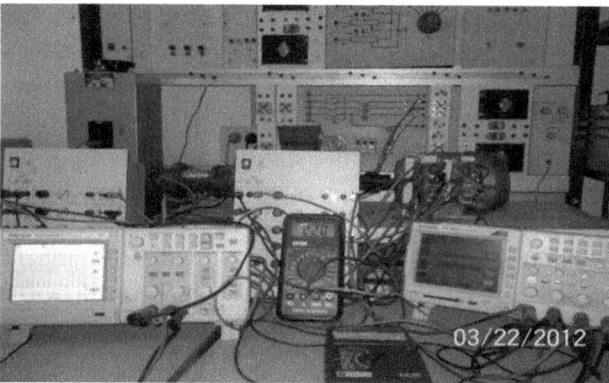

Fig. 1. Test bench.

Figure 2 shows the development panoramic coil of two stator windings of the learning machine, which are spread over 24 slots. Frontal connections of winding 1 of phase 1 are heavy solid line, those of phase two and phase three are respectively strong dotted then strongly mixed lines. Concerning frontal connections of winding 2 of phases 1, 2 and 3 are in continuous, dotted and mixed lines. The angle θ_{mec} represents the mechanical shift between the two stars, which belongs to the set $\{0/P, 30/P, 60/P\}$ with P is the number of pole pairs.

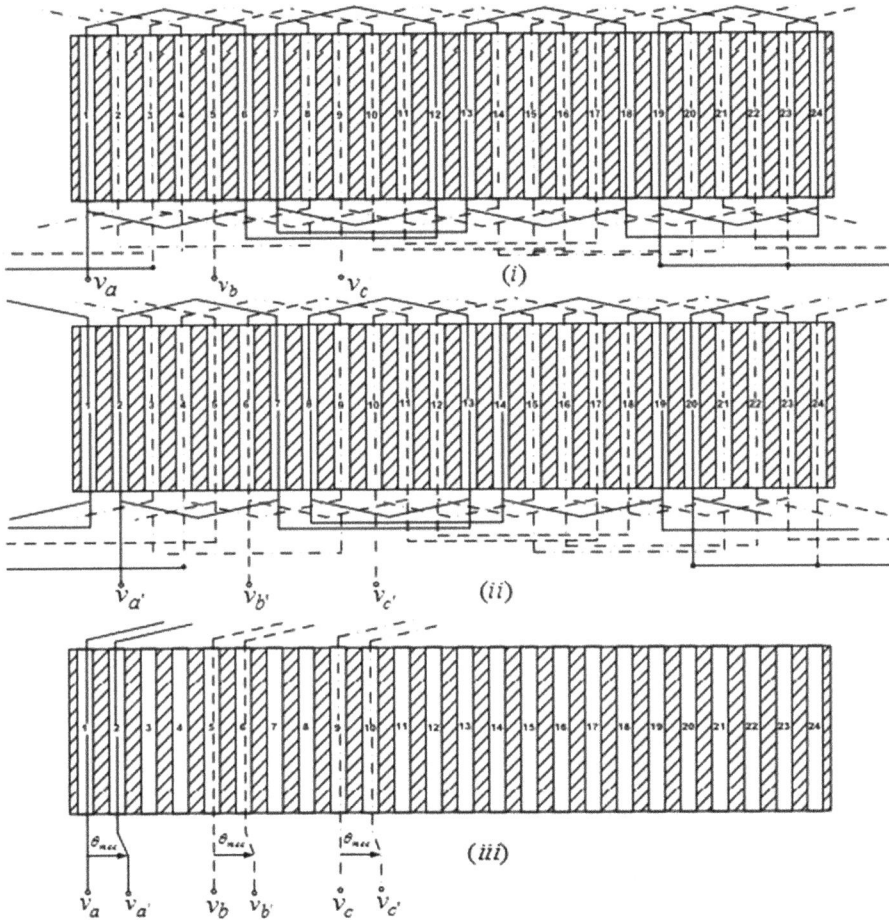

Fig. 2. Panoramic development of stator windings of DSIM, 0.5 kW, 4 poles, 24 slots and two beams by slot. (i): winding 1. (ii): winding 2. (iii): mechanical shift between the two stars.

3 Transient model of double star induction machine

As shown in Fig. 3, the machine has two stator windings sets (a, b, c) and (a', b', c') spatially shifted by α, with isolated neutral points and an equivalent three-phase squirrel-cage rotor. Angles θ' and $(\theta' - \alpha)$ denote the rotor position respectively to star 1 and 2.

For this machine, we adopt the following assumptions:
-   Stator windings are sinusoidally distributed,
-   windings are identical within each three phase set.

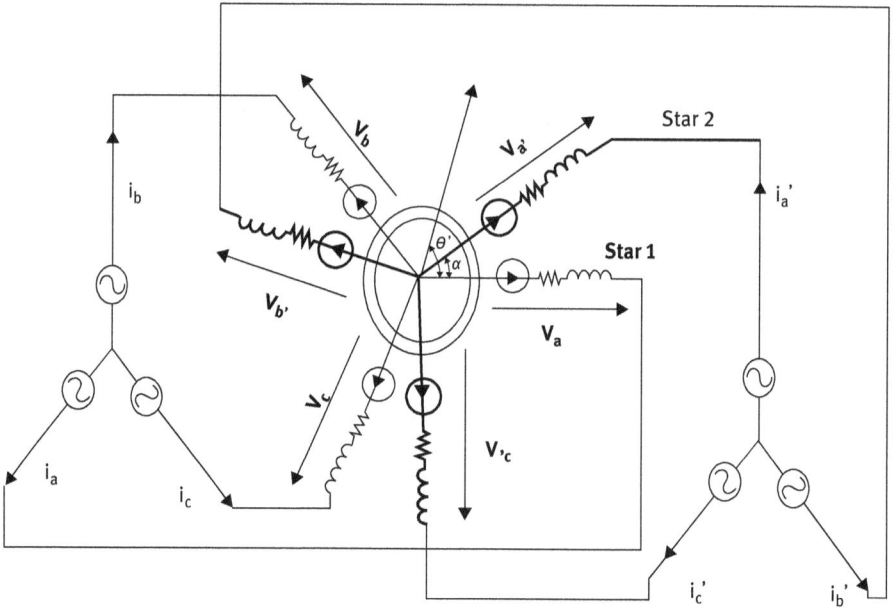

Fig. 3. Dual star induction machine.

Characteristic equations of a DSIM in a common reference frame are:

$$
\begin{cases}
\overline{V}_{s1} &= R_s \overline{i}_{s1} + \dfrac{d\overline{\lambda}_{s1}}{dt} + w_a \overline{\lambda}_{s1} \\[2mm]
\overline{V}_{s2} &= R_s \overline{i}_{s2} + \dfrac{d\overline{\lambda}_{s2}}{dt} + w_a \overline{\lambda}_{s2} \\[2mm]
0 &= R_r \overline{i}_r + \dfrac{d\overline{\lambda}_r}{dt} + (w_a - w_m)\overline{\lambda}_r
\end{cases}
\tag{1}
$$

with:

$$
\frac{J}{P} \times \frac{dw_m}{dt} = T_{em} - T_{load}
\tag{2}
$$

$\omega_a = \dfrac{d\theta_a}{dt}$ is the reference speed, and $\omega_m = \dfrac{d\theta'}{dt}$ is the electrical rotor speed.
Flux expressions are:

$$\begin{cases} \bar{\lambda}_{s1} &=& (l_s + l_{sm})\bar{i}_{s1} + l_{sm}\bar{i}_{s2} + \bar{\lambda}_m \\ \bar{\lambda}_{s2} &=& l_{sm}\bar{i}_{s1} + (l_s + l_{sm})\bar{i}_{s2} + \bar{\lambda}_m \\ \bar{\lambda}_r &=& l_r\bar{i}_r + \bar{\lambda}_m \end{cases} \qquad (3)$$

where:

$$\bar{\lambda}_m = L_m\bar{i}_m \qquad (4)$$
$$\bar{i}_m = \bar{i}_{s1} + \bar{i}_{s2} + \bar{i}_r \qquad (5)$$

As shown in Fig. 4 and Fig. 5, in the electric equivalent circuit of DSIM, the stator mutual leakage inductance is included in the common branch of the two stars [19]. It is more convenient to separate the space vector equations in $d-q$ ones. The d–axis equations are:

$$\begin{cases} \bar{\lambda}_{ds1} &=& (l_s + l_{sm})\bar{i}_{ds1} + l_{sm}\bar{i}_{ds2} + L_m\bar{i}_{dm} \\ \bar{\lambda}_{ds2} &=& l_{sm}\bar{i}_{ds2} + (l_s + l_{sm})\bar{i}_{ds2} + L_m\bar{i}_{dm} \\ \bar{\lambda}_{dr} &=& l_r\bar{i}_{dr} + L_m\bar{i}_{dm} \end{cases} \qquad (6)$$

and the q–axis equations are:

$$\begin{cases} \bar{\lambda}_{qs1} &=& (l_s + l_{sm})\bar{i}_{qs1} + l_{sm}\bar{i}_{qs2} + L_m\bar{i}_{qm} \\ \bar{\lambda}_{qs2} &=& l_{sm}\bar{i}_{qs2} + (l_s + l_{sm})\bar{i}_{qs2} + L_m\bar{i}_{qm} \\ \bar{\lambda}_{qr} &=& l_r\bar{i}_{qr} + L_m\bar{i}_{qm} \end{cases} \qquad (7)$$

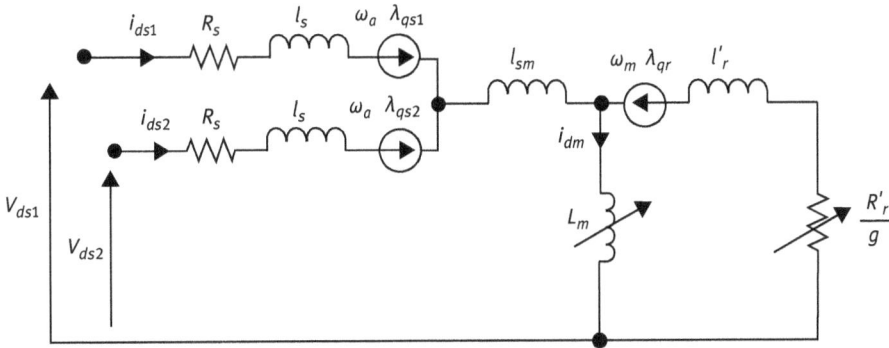

Fig. 4. d-axis equivalent circuit of DSIM in an arbitrary reference frame.

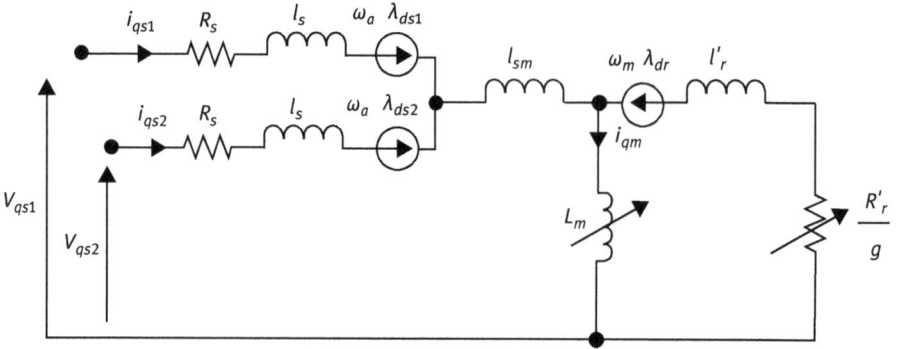

Fig. 5. q-axis equivalent circuit of DSIM in an arbitrary reference frame.

4 Introduction of magnetic saturation effects

The adopted philosophy considers the self flux of a coil as the superposition of a leakage's flux and a useful flux traversing the iron. As a consequence, leakage inductances become constant. This implies that only the main flux is subject to magnetic saturation.

Static and dynamic mutual inductances are defined respectively, as follows:

$$
\begin{cases}
L_m & = \dfrac{\lambda_m}{i_m} \\[2mm]
L_{mdy} & = \dfrac{d\lambda_m}{di_m}
\end{cases}
\tag{8}
$$

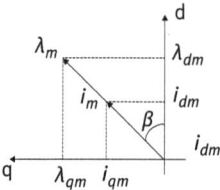

Fig. 6. Flux and current magnetizing.

According to Fig. 6, the $d-q$ components of the magnetizing flux and current are:

$$
\begin{cases}
\lambda_{md} & = \lambda_m \cos\beta \\
\lambda_{mq} & = \lambda_m \sin\beta
\end{cases}
\qquad
\begin{cases}
i_{md} & = i_m \cos\beta \\
i_{mq} & = i_m \sin\beta
\end{cases}
\tag{9}
$$

with:

$$
L_m = \frac{\lambda_m}{i_m} = \frac{\lambda_{md}}{i_{md}} = \frac{\lambda_{mq}}{i_{mq}}
\tag{10}
$$

For modeling a saturated squirrel cage double star induction machine in $d - q$ axis, the proposed method is based mainly on the winding model of the current. Since $(i_{ds1}, i_{qs1}, i_{ds2}, i_{qs2}, i_{dm}, i_{qm})$ constitute the selected state variables, the rotor flux and current and their time derivatives in the primitive d-q set of Equation (1), must be written as their functions. The $d - q$ components of $(\lambda_{s1}, \lambda_{s2}, \lambda_r)$ are normally written in terms of the winding currents in the system of Equations (6) and (7). Derived stator and rotor linkage fluxes in Equations 6 and 7 leads to the time derivative of the magnetizing inductance L_m.

The leakage inductances in Equation (3) are assumed to be constant, only the main flux $\overline{\lambda}_m$ is subject to saturation. Thus, $\dfrac{d\lambda_m}{dt}$ has to be described by means of the winding currents.

Now let's write:

$$\frac{dL_m}{dt} = \frac{dL_m}{di_m} \times \frac{di_m}{dt} \tag{11}$$

From Equation (4), $L_m = \frac{\lambda_m}{i_m}$ and after deriving $i_m = \sqrt{i_{dm}^2 + i_{qm}^2}$, we get, respectively (12) and (13):

$$\frac{dL_m}{di_m} = \frac{L_{mdy} - L_m}{i_m} \tag{12}$$

$$\frac{di_m}{dt} = \frac{i_{dm}}{i_m} \times \frac{di_{dm}}{dt} + \frac{i_{qm}}{i_m} \times \frac{di_{qm}}{dt}$$

$$= \cos\beta \frac{di_{dm}}{dt} + \sin\beta \frac{di_{qm}}{dt}$$

$$= \cos\beta \frac{di_{ds1}}{dt} + \sin\beta \frac{di_{qs1}}{dt} + \cos\beta \frac{di_{ds2}}{dt} + \sin\beta \frac{di_{qs2}}{dt}$$

$$+ \cos\beta \frac{di_{dr}}{dt} + \sin\beta \frac{di_{qr}}{dt} \tag{13}$$

Argument β is the angular position of space vector $\bar{i}_m$ (or $\overline{\lambda}_m$) with respect to d-axis so that:

$$\beta = \arctan \frac{i_{qm}}{i_{dm}} \tag{14}$$

Finally, we have:

$$\frac{d\lambda_{ds1}}{dt} = (l_s + l_{sm})\frac{di_{ds1}}{dt} + l_{sm}\frac{di_{ds2}}{dt} + L_d\frac{di_{dm}}{dt} + L_{dq}\frac{di_{qm}}{dt} \tag{15}$$

$$\frac{d\lambda_{qs1}}{dt} = (l_s + l_{sm})\frac{di_{qs1}}{dt} + l_{sm}\frac{di_{qs2}}{dt} + L_{dq}\frac{di_{qm}}{dt} + L_q\frac{di_{dm}}{dt} \tag{16}$$

$$\frac{d\lambda_{ds2}}{dt} = l_{sm}\frac{di_{ds1}}{dt} + (l_s + l_{sm})\frac{di_{ds2}}{dt} + L_d\frac{di_{dm}}{dt} + L_{dq}\frac{di_{qm}}{dt} \tag{17}$$

$$\frac{d\lambda_{qs2}}{dt} = l_{sm}\frac{di_{qs1}}{dt} + (l_s + l_{sm})\frac{di_{qs2}}{dt} + L_{dq}\frac{di_{qm}}{dt} + L_q\frac{di_{dm}}{dt} \tag{18}$$

$$\frac{d\lambda_{dr}}{dt} = -l_r\frac{di_{ds1}}{dt} - l_r\frac{di_{ds2}}{dt} + (l_r + L_d)\frac{di_{dm}}{dt} + L_{dq}\frac{di_{qm}}{dt} \tag{19}$$

$$\frac{d\lambda_{qr}}{dt} = -l_r\frac{di_{qs1}}{dt} - l_r\frac{di_{qs2}}{dt} + L_{dq}\frac{di_{dm}}{dt} + (l_r + L_q)\frac{di_{qm}}{dt} \tag{20}$$

with:

$$L_{dq} = (L_{mdy} - L_m)\cos\beta\sin\beta \tag{21}$$

$$L_d = L_{mdy} - (L_{mdy} - L_m)\sin^2\beta \tag{22}$$

$$L_q = L_{mdy} - (L_{mdy} - L_m)\cos^2\beta \tag{23}$$

Using the equation $V = A\dot{X} + BX$, with X is a vector formed by $d-q$ components of the winding currents and $\dot{X}$ its time derivative, matrices A and B are:

$$A = \begin{bmatrix} l_s + l_{sm} & 0 & l_{sm} & 0 & L_d & L_{dq} \\ 0 & l_s + l_{sm} & 0 & l_{sm} & L_{dq} & L_q \\ l_{sm} & 0 & l_s + l_{sm} & 0 & L_d & L_{dq} \\ 0 & l_{sm} & 0 & l_s + l_{sm} & L_{dq} & L_q \\ -l_r & 0 & -l_r & 0 & l_r + L_d & L_{dq} \\ 0 & -l_r & 0 & -l_r & L_{dq} & l_r + L_q \end{bmatrix} \tag{24}$$

$$B = \begin{bmatrix} R_s & -c_1 & 0 & -d_1 & 0 & -\omega_a L_m \\ c_1 & R_s & d_1 & 0 & \omega_a L_m & 0 \\ 0 & -d_1 & R_s & -c_1 & 0 & -\omega_a L_m \\ d_1 & 0 & c_1 & R_s & \omega_a L_m & 0 \\ -R_r & a_1 & -R_r & a_1 & R_r & -b_1 \\ -a_1 & -R_r & -a_1 & -R_r & b_1 & R_r \end{bmatrix} \tag{25}$$

with $a_1 = (\omega_a - \omega_m)l_r$, $b_1 = (\omega_a - \omega_m)(l_r + L_m)$, $c_1 = \omega_a(l_s + l_{sm})$ and $d_1 = \omega_a l_{sm}$.

We can notice that the elements of matrix A depend on the saturation. Moreover, it contains all kinds of magnetic coupling along the d–axis (L_d), q–axis (L_q), and the $d-q$ axis (L_{dq}).

5 Introduction of skin effect

Rotor resistance varies not only by thermal effects, but also by the skin effect. Indeed, skin effect effectively increases rotor resistance and decreases rotor leakage inductance. Therefore, in order to take into account the skin effect we generally use the following linear equations:

$$R_r = R_{rdc} + (R_{rl} - R_{rdc})\sqrt{\frac{s - s_n}{1 - s_n}} \tag{26}$$

$$X_r = X_{rdc} + (X_{rl} - X_{rdc}) \left(\frac{s - s_n}{1 - s_n} \right)^2 \qquad (27)$$

where: R_{rl}: rotor resistance at 50 Hz, determined from rotor-blocked test. R_{rdc}: rotor resistance at zero (dc value) X_{rl}: rotor leakage reactance at 50 Hz, determined from rotor-blocked test. X_{rdc}: rotor leakage reactance at synchronous speed.

6 Experimental and simulation results

In order to validate the developed dynamic model of the double star induction machine, an experimental study was carried out by using a 6-phases, 500W, 220V/380V; 4-poles. The machine is supplied directly from the power distribution network.

Figure 7 gives a comparison between the measured and simulated torque values at run-up from rest. The torque peak- values computed from model with saturation and skin effect are higher than computed values using conventional model. Moreover, the time taken to reach the steady state for accurate model is very close to that of experimental results.

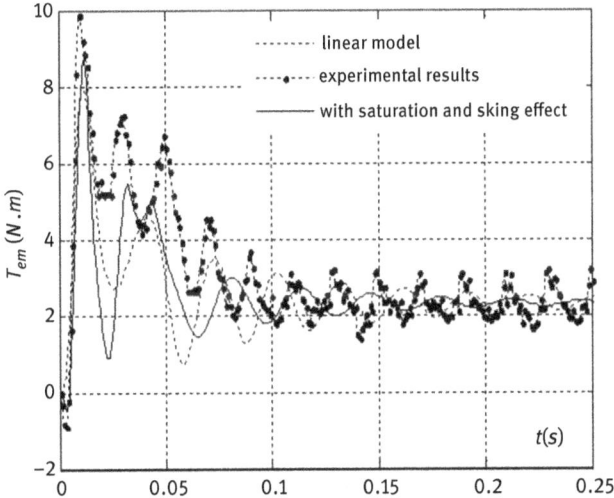

Fig. 7. Comparison between measured and simulated transient torque during run-up.

Furthermore, Fig. 8 shows the comparison between simulation and experimental results of rotor speed in transient regime at start-up on load operation. We also note that at start-up, if the saturation and the skin effect are taken into consideration, the rotor speed of the DSIM is closer to the measured rotor speed than it is if they are ignored.

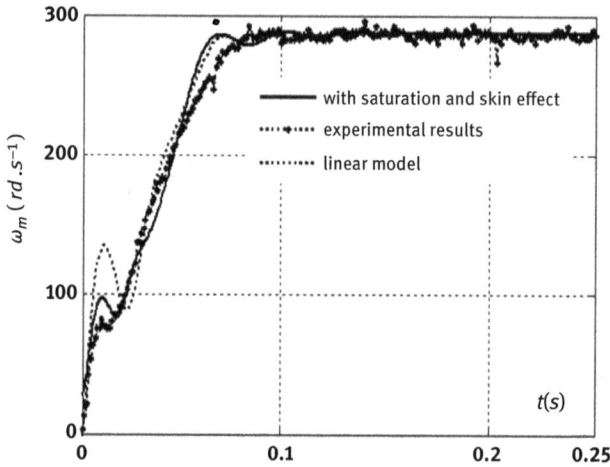

Fig. 8. Comparison between measured and simulated transient rotor speed during run-up.

From the mechanical characteristics, we can confirm that the impact of the saturation is clearer in transient regime and especially at start-up.

Figure 9 illustrates the comparison between simulation and experimental results of the stator current of DSIM for the same operation. It can be observed that there's

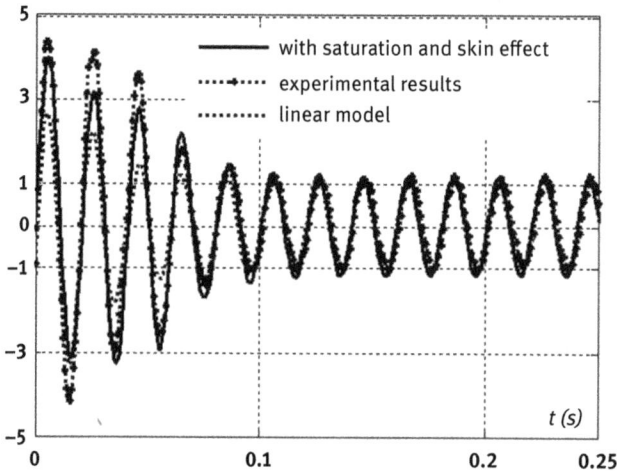

Fig. 9. Stator current of DSIM during run-up.

a little difference at the first peak value. Although this difference appears, the saturation model still gives more accurate results for the stator current than it is for the conventional machine model. However, skin effect phenomena induce variation in rotor parameters. Considering the skin effect into rotor parameters, this difference can be slightly reduced.

We note that taking into account the magnetic saturation and skin effect improves the accuracy of predicting the performance of the machine in transient state. However, there are still differences between the results of simulations and measurements especially in the early peaks. This can be explained by the fact that we have ignored the effect of saturation in the leakage flux.

7 Conclusion

A dynamic model of the double star induction machine that takes into account the magnetic saturation, the stator mutual leakage inductance between two stars and the skin effect is proposed in this paper. A comparison between simulation and experimental results shows that the proposed model is closer to measurement results than in the linear model, this, is validated for mechanical and electrical characteristics of the machine. It is shown that if we take into account the magnetic saturation and the skin effect, this will improve the accuracy and the performances of the prediction of the double star induction machine and particularly in transient operation and at startup.

Bibliography

[1] Farag K. Abo-Elyousr and G. H. Rim. Performance Evaluation of AC/DC PWM Converter for 12-phase Stand-Alone PMSG with Maximum Power Extraction. *Int. Conf. on Electrical Machines and Systems* (ICEMS), Incheon, :82–87, December 2010.

[2] J.M. Apsley, S. Williamson, A.C. Smith and M. Barnes. Induction motor performance as a function of phase number. *IEE Proc. Electric Power Applications*, 153(6):898–904, November 2006.

[3] Z. Oudjebour. Stabilization by New control technique of the input DC voltages of five-level diode - Clamped inverters. Application to double star induction machine. *2nd Int. Symp. on Environment Friendly Energies and Applications* (EFEA), 2012, :541–544.

[4] A. Tessarolo. Benefits of increasing the number of stator phases in terms of winding construction technology in high- power electric machines. *5th IET Int. Conf. on Power Electronics, Machines and Drives* (PEMD), :1-6, 2010.

[5] S. Williamson and S. Smith. Fault tolerance in multiphase propulsion motors. *J. of Marine Engineering and Technology*, 4(5):3–7, March 2004.

[6]   G. K. Singh and V. Pant. Analysis of multiphase induction machine under fault condition in a phase-redundant ac drive system. *Int. J. Electric Machines Power Systems*, 28(6):577–590, June 2000.
[7]   S. Guizani, F. Ben Ammar. The eigenvalues analysis of the double star induction machine supplied by redundant voltage source inverter. *Int. Review of Electrical Engineering* (I.R.E.E), 3(2), April 2008.
[8]   M.T. Mohammad and J.E. Fletcher. Five-phase Permanent Magnet Machines, Advantages and applications. 5th *IET Int. Conf. on Power Electronics, Machines and Drives* (PEMD), :1–5, 2010.
[9]   A. Khedher and M. F. Mimouni. Sensorless-adaptive DTC of double star induction motor. *Energy Conversion and Management*, 51(12):2878–2892, December 2010.
[10] H. Rehaoulia, H. Henao and H. Capolino. G.A.: Modeling of synchronous machines with magnetic saturation. *Electric Power Systems Research*, 77(5):652–659, July 2007.
[11] Tu. Xiaoping, A. Louis Dessaint, Roger Champagne and K. Al-Haddad. Transient Modeling of Squirrel-Cage Induction Machine Considering Air-Gap Flux Saturation Harmonics. *IEEE Trans. on Industrial. Electronics*, 55(7):2798–2809, July 2008.
[12] G. K. Singh. Multi-phase induction machine drive research - a survey. *Electric Power Systems Research*, 61(2):139–147, March 2002.
[13] E. Levi. Multiphase electric machine for variable speed applications. *IEEE Trans. Industrial. Electronics*, 55(5):1893–1909, May 2008.
[14] T. A. Lipo. A d-q Model for Six Phase Induction Machines. *Int. Conf. on Electrical Machines*, :860–867, 1980.
[15] H. Razik. Modelling of double star induction motor for diagnosis purpose. *Conf. on Electric Machines and Drives*, :907–912, 2003.
[16] R. H. Nelson and P. C. Krause. Induction machine analysis for arbitrary displacement between multiple winding sets. *IEEE Trans. Power Apparatus Systems*, 93(3):841–848, 1974.
[17] V. Pant, G.K. Singh and S.N. Singh. Modeling of a multi-phase induction machine under fault condition. *IEEE Int. Conf. Power Electrical Drives*, :92–97, Hong Kong, 1999.
[18] N. Erdogan, T. Assaf, R. Grise and M. Aubouroug. An Accurate 3-phase Induction Machine Model Including Skin Effect and Saturation for Transient Studies. *Electrical Machines and Systems*, ICEMS 2003, 2:646–649.
[19] H. Kouki, M. Ben Fredj and H. Rehaoulia. Effect of the stator mutual leakage inductance for low and high power applications of dual star induction machine. *Int. Review on Modelling and Simulations* (I.RE.MO.S.), 5(2), April 2012.

Biographies

Hajer Kouki received the B.S. degree in Electrical Engineering and Power Electronics from the ESSTT (Institute of Sciences and Technology) in Tunis, Tunisia, in 2005, the M.S. degree in electric systems from the ENIT (National Institute of Engineering of Tunis), in 2007. Since 2009, she's an Assistant in ISSTEG in the Department of Electrical Engineering. Her main research interests are the modeling, simulation, of electrical machines and power electronics.

Mouldi Ben Fredj received the master's degree in electrical and electronic automatic in 1985 and the doctoral degree in april 1989, from the University of Science and Technology of Lille France. Since 2000 he is with the ESSTT (Institute of Sciences and Technology of Tunis). His main research interests are power electronics and control of electrical machines: modeling and simulation of association's electrical machines -static converters, renewable energy sources.

Habib Rehaoulia is a Professor of electrical engineering. He received the B.E. degree in 1978, the M.S. degree in 1980, the doctoral degree in 1983 and the habilitation degree in 2007, all from the ENSIT (National high school of engineers), University of Tunis, Tunisia. He joined the teaching staff of the ENSET (lately ENSIT) in 1978. During his career, he was on leave for several months at WEMPEC (University of Madison Wisconsin USA), ENSIEG (University of Grenoble France), "Electrotechnic Lab." (University of Paris VI France), and CREA (University of Picardie France). His main research interests are analysis and modeling of electrical machines.

M. Lindner, P. Braeuer and R. Werner

Increasing the Torque Density of Permanent-Magnet Synchronous Machines using Innovative Materials and Winding Technologies

Abstract: Due to the progress of system integration, the increase of energy efficiency and device mobility it is worthwhile to investigate alternative materials for electrical machines and methods of producing electric windings. In this article the influence of innovative materials on the machine behaviour as well as the machine design are discussed. This includes **FeCo**, **FeNi** and **SMC**. For each of these materials design guidelines are defined. Beyond this the screen printing technology – so far used since a couple of years especially for producing cheap windings for small-power electrical motors – is generalized to a wide range of machine sizes. A possible benefit over classic copper wire windings regarding torque density is determined.

Keywords: screen-printing, soft-magnetic material, iron-cobalt, SMC, utilization, electric machines, manufacturing.

Mathematics Subject Classification 2010: 65C05, 62M20, 93E11, 62F15, 86A22

1 Introduction

With electrical machines penetrating increasingly more domains of everyday life new demands have been raised for them. Drives for handheld devices have to be sized as small as possible to achieve a high mobility or some flexibility to distribute the mass. Examples are hand tools for craftsmen, medical technology or household purposes. Even more important is to reduce the weight of moving motors. Lower driven masses require lower acceleration torque and hence yet lower motor mass. [1] Furthermore, the system efficiency can be increased. The main focus in this respect is the transportation sector. Electrical and hybrid cars, space- and aircrafts but even conventional cars with hundreds of auxiliary actuators. [2] Further examples are conveying systems, stair lifts and robots.

Conventional machine design technology limits the torque density, described by the utilization factor $C = T/V$ with torque T and machine volume V, to a value of approx. $10\,\mathrm{kN\,m^{-2}}$ [3] for air-cooled permanent magnet synchronous motors (PMSM). This magnitude results from a maximum LORENTZ force product – air gap flux density

M. Lindner, P. Braeuer and R. Werner: Chair of Electrical Energy Conversion Systems and Drives, Chemnitz University of Technology, 09107 Chemnitz, Germany, ewainfo@etit.tu-chemnitz.de

De Gruyter Oldenbourg, ASSD – Advances in Systems, Signals and Devices, Volume 3, 2017, pp. 97–113.
DOI 10.1515/9783110448412-007

B_g times orthogonal stator current I_{sq}. The achievable flux density deeply depends on the material and configuration of the magnets. However, another great influence is exerted by the magnetization curve of the soft magnetic material and the width of flux conducting sections as teeth and yoke. The current limit depends on the maximum current density of the winding, hence it's resistance and surface area, as well as the cross-section of the slots. The latter highly interrelates to the previously mentioned width of soft magnetic sections with radial flux machines. Furthermore, to guarantee the winding temperature the tolerable current density depends on the electric loading via the motor's electrical stress, which is defined e. g. by the magnetic core loss and the thermal conductivity of the core. A magnitude rather influencing the volume is the geometry of the end winding, which produces unprofitable room decreasing the utilization factor. Summarized, conventional machine designs are mainly limited by the material properties and production constraints of the soft magnetic core and the windings.

One possibility to avoid those constraints is to adopt alternative core materials, which will be investigated in section 2. A second approach directly deduced from the previous analysis utilizes a novel winding technology, which is described in section 3. The achievable improvements with both proposals are summarized in section 4.

All calculations were performed analytically in order to achieve general predictions as well as direct knowledge about the influences of single parameters instead of special solutions numerical calculations deliver. Furthermore, well proven analytical models provide a quality regarding overall machine predictions as good as numerical approaches do – merely local values differ in higher quantity.

2 Influence of Innovative Materials

2.1 Material Selection

Soft magnetic materials differ – beside other physical and mechanical properties – in their magnetic polarization J and specific loss p. These parameters strongly determine the behaviour of the electrical machine. The higher the polarization the smaller the required soft magnetic cross section with a constant magnetic flux. This leads to a smaller overall machine volume or more design space for nonferrous sections as slots or permanent magnets. Furthermore, the lower the specific loss the higher the machine efficiency and the lower the rated temperature rise.

The standard ferromagnetic material in electrical machine construction are sheets from an iron-silicon alloy (FeSi). [4] A much higher saturation polarization and lower specific loss can be achieved with iron-cobalt (FeCo). Alloys from iron-nickel (FeNi) have even lower loss but also suffer from lower saturation. In addition to the alloy different delivery forms and production processes can be considered. Soft magnetic

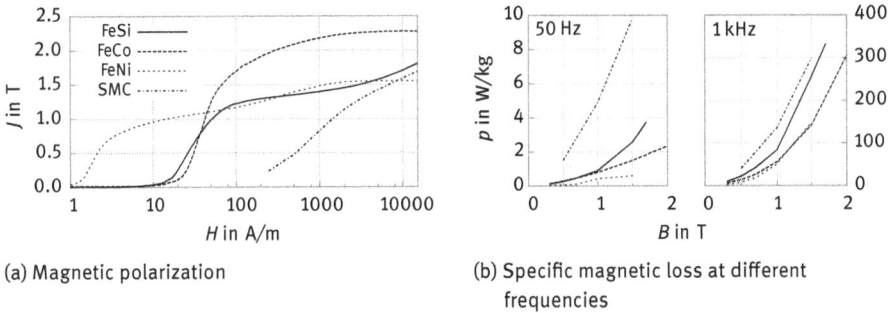

(a) Magnetic polarization

(b) Specific magnetic loss at different frequencies

Fig. 1. Characteristics of the considered materials. **FeSi**: TRAFOPERM N4; **FeCo**: VACOFLUX 50; **FeNi**: PERMENORM 5000V5 (all VACUUMSCHMELZE GMBH & CO. KG [7]); **SMC**: SOMALOY 700 3P (HÖGANÄS AB [8]).

composites (SMC) are powders typically made from pure iron [5]. They allow less eddy currents but cause higher hysteresis loss and lower saturation compared to sheets. Further details on those material properties and their origin can be found in [6]. Fig. 1 compares the described characteristics of magnetic polarization and specific loss based on representative grades.

A superior behaviour show nanocrystalline materials. They have high saturation polarizations and very low specific loss. However, currently, they can only be produced in thin and brittle ribbons hardly applicable to machine construction. Therefore, they will not be considered consecutively.

2.2 Effects on the machine design

The determination of materials' effects on the utilization factor of PMSM is deduced from [3] and [9] and was further enhanced. In oder to reasonably compare the results minimizing the influence of disregarded parameters, as e. g. stray, several predefinitions need to be established:

(PD 1) The main geometries and parameters as outer machine diameter and length, inner diameter, the rotor as well as the number of slots, poles and windings remain unchanged.

(PD 2) The air gap is the highest magnetic resistance in the magnetic circuit.

(PD 3) From (PD 1) and (PD 2) follows that the permanent magnets remain in the same operation point. To support this assumption the magnetic field is kept constant in the stator sections of the magnetic circuit.

(PD 4) From (PD 1) and (PD 3) follows that the magnetic flux and induction in the air gap are constant.

(PD 5) The tooth flanks are parallel.
(PD 6) Armature reaction, thus effects of the stator on the rotor, are neglected.
(PD 7) The rated temperature rise of the winding is constant.

The calculation method discussed consecutively works with normalized machines. Instead of establishing precise numerical quantities the coefficients of modified values are determined. Thus, the results are transferable to most machine sizes. Fig. 2 shows the applied calculation sequence.

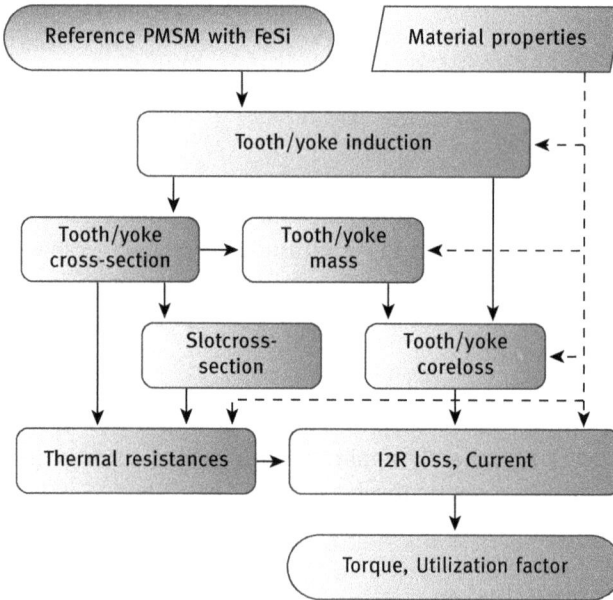

Fig. 2. Concept chart of calculating the change of utilization factor with new ferromagnetic materials.

A magnetization curve as seen in Fig. 1a allows to directly gain the ratio of magnetic inductions (B/B_{FeSi}) at constant field strength according to (PD 3). This can be done both for the stator tooth and yoke. With constant air gap flux Φ_g, (PD 4), the cross-sections have to change consequently.

$$\frac{A}{A_{\text{FeSi}}} = \frac{\Phi_g}{B} \cdot \frac{B_{\text{FeSi}}}{\Phi_g} = \left(\frac{B}{B_{\text{FeSi}}}\right)^{-1} \quad \text{at} \quad H = const. \tag{1}$$

This directly applies to the ratio of tooth widths $(w_t/w_{t,\text{FeSi}})$ and yoke heights $(h_y/h_{y,\text{FeSi}})$ with constant machine length according to (PD 1). Thus, the stator mass changes considering additionally the varying mass density of the materials.

With (PD 1) changing the tooth width and yoke height modifies the slot width w_s and height h_s and hence the cross-section of copper A_{Cu} at constant slot fill factor, which is inverse to the ohmic resistance of the stator winding R_s. With a trapezoidal slot this is:

$$\left(\frac{R_s}{R_{s,\text{FeSi}}}\right)^{-1} = \frac{A_{Cu}}{A_{Cu,\text{Fesi}}} \approx \frac{A_s}{A_{s,\text{Fesi}}} = \frac{w_s}{w_{s,\text{FeSi}}} \cdot \frac{h_s}{h_{s,\text{FeSi}}}. \tag{2}$$

By means of the loss data in Fig. 1(b) ratios of the specific core loss for the new material $(P_c/P_{c,\text{FeSi}})$ can be achieved taking into account the change of induction in tooth and yoke.

Furthermore, with different geometries and material-specific thermal conductivities the thermal resistances of the stator sections change, e. g. of the yoke above the slot:

$$\frac{R_{th\,ys}}{R_{th\,ys,\text{FeSi}}} = \frac{h_y}{h_{y,\text{FeSi}}} \cdot \left(\frac{\lambda_{th}}{\lambda_{th,\text{FeSi}}} \cdot \frac{w_s}{w_{s,\text{FeSi}}}\right)^{-1}. \tag{3}$$

The thermal behaviour of the considered machines can be estimated with the lumped parameter thermal network in Fig. 3. In doing so the heat flow across the air gap and in axial direction is neglected for the benefit of light-weight relationships. This is reasonable because of few expected influences due to (PD 1), a high thermal resistivity of air and the axially consistent structure. [10]

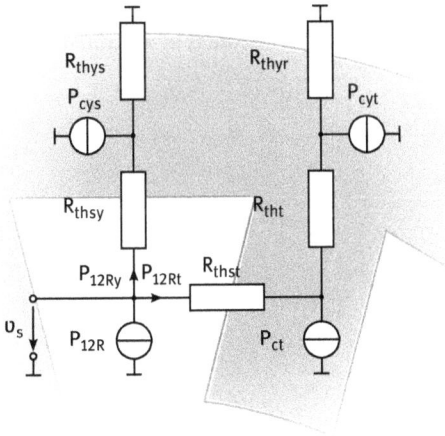

Fig. 3. Simple lumped parameter thermal network.

The keynote to thermal considerations is based on (PD 7), thus a constant slot temperature with changing core power losses and thermal resistances. Since the same slot materials are supposed a maximum insulation temperature is crucial to machine design. With the assumptions of (a) homogeneous loss distribution in the core sections and (b) equal specific heat transfer through slot side and bottom a circuit analysis leads to the ratio of I2R loss ($P_{I2R}/P_{I2R, FeSi}$). The thermal resistance of the slot was estimated for enamel-insulated round wires with rated diameters of 1 mm. [11, 12]

Finally, the new phase current I_{ph} results from the change of ohmic loss and resistance. With (PD 4) the torque T scales proportionally. The same applies for the utilization factor C with (PD 1).

$$\frac{C}{C_{FeSi}} = \frac{T}{T_{FeSi}} = \frac{I_{ph}}{I_{ph, FeSi}} = \sqrt{\frac{P_{I2R}}{P_{I2R, FeSi}} \cdot \left(\frac{R_s}{R_{s, FeSi}}\right)^{-1}} \tag{4}$$

Most machine parameters can be directly calculated. However, the determination of utilization variation requires assessing three quantities. The frequency f as well as the field strengths in yoke H_y and tooth H_t have to be set in order to identify changes in the material characteristics Fig. 1. Furthermore seven ratios of the reference FeSi machine are required. The tooth to slot width ($w_{t, FeSi}/w_{s, FeSi}$), yoke to slot height ($h_{y, FeSi}/h_{s, FeSi}$) and yoke height to yoke diameter ($h_{y, FeSi}/d_{y, FeSi}$) help to define the change of slot area. Additionally, both the tooth core loss to I2R loss ($P_{ct, FeSi}/P_{I2R, FeSi}$), the yoke core loss to I2R loss ($P_{cy, FeSi}/P_{I2R, FeSi}$), the I2R loss dissipating over the tooth flank to overall I2R loss ($P_{I2R t, FeSi}/P_{I2R, FeSi}$) and the slot height to slot width ($h_{s, FeSi}/w_{s, FeSi}$) are needed to estimate the change of I2R loss. All those parameters were varied reasonably in order to find dependencies of the utilization factor. The examined ranges are shown in Tab. 1. Finally, rules of thumb about the optimal usage as well as the expected changes are developed and mentioned in section 4.

Tab. 1. Variation of reference machine parameters.

	min	max
f/Hz	50	1000
H_t/Am^{-1}	500	10,000
H_y/Am^{-1}	500	10,000
$w_{t, FeSi}/w_{s, FeSi}$	0.5	1.5
$h_{y, FeSi}/h_{s, FeSi}$	0.5	1.5
$h_{y, FeSi}/d_{y, FeSi}$	0.03	0.075
$h_{s, FeSi}/w_{s, FeSi}$	0.5	1.5
$P_{ct, FeSi}/P_{I2R, FeSi}$	0.1	1.0
$P_{cy, FeSi}/P_{I2R, FeSi}$	0.1	1.0
$P_{I2R t, FeSi}/P_{I2R, FeSi}$	0.3	0.9

3 Influence of Innovative Winding Technologies

3.1 Conventional Windings

Beside new materials also new technologies especially for the design of the winding can be useful to increase the torque density of electrical machines. In classic terms electric motors and generators with windings comprised of copper or aluminium are known. These are produced in a wide range of variants. Especially energy converters with distinct phase windings, commutator windings, windings with distinct poles and tooth coils should be mentioned [9], [13]. The winding is typically produced by coiling isolated copper wire. Depending on the type of winding, the process of coiling is one of the most expensive steps during fabrication of electrical machines. Furthermore, beside permanent magnets the winding is the temperature limiting element of an motor because of its isolation.

3.2 Screen Printed Windings

The screen printing process has its origin in the printing industry. Since the 1960s it is used for printing conductive, insulating and resistive layers in the electronics industry [14]. The development in the area of printing forms as well as the availability of new material systems and more precise screen printing machines in the last years are the reason for distinct better producible layers and structures.

Actually, screen printing is used for printing aerials for RFID smart labels [15]. Since at least two years it is possible to utilize this procedure in drive engineering. Thus, a reduction of the layer thickness of the winding as well as a reduction of costs are conceivable.

The first sample of windings for electrical energy converters printed at the Institute for Print and Media Technology at Chemnitz University of Technology (pmTUC) and tested in our laboratory was a threephase stator winding for a permanent magnetic synchronous motor described in [16] and [17]. Its excitation is done by a rotor made of NdFeB magnets. The layout for this machine consists of two conductive layers printed with silver paste and two insulating layers printed with a dielectric paste. Both conductive silver layers are connected through an interlayer connection. In Fig. 4 an example of a screen printed winding is shown. Currently, the windings were printed with a semi automatic screen printing machine with an optical positioning system. A PET plastic film with a thickness of 50 µm is used as substrate but also PEN and PEEK or ceramic foils and smaller thicknesses are conceivable. The drying of the printed layers has taken place at a temperature of 150 °C. The screen used has a cloth of 120 threads per centimetre with a denier of 34 µm. Its mesh size is 45 µm and the rake angle 22.5 °. As substrate DuPont Teijin Melinex 401, a 50 µm PET plastic film, was used. All layers were printed with a print speed of 100 mm/s and a squeegee pressure of 1 bar. For the conductive layers the silver paste DuPont 5029 was used and dried for 5 min at 120 °C with a belt drier. All dielectric layers were printed with DuPont 5018G and dried with UV light at approximately 6.5 J/cm^2.

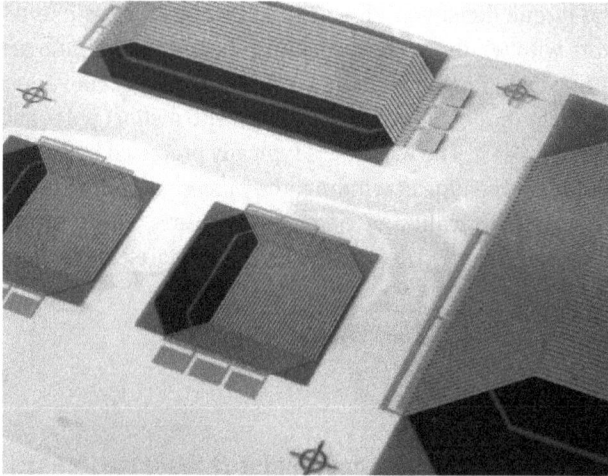

Fig. 4. Samples of screen printed air-gap-windings of different three phase synchronous small-power electrical machine.

3.3 Effects on the machine design

Used instead of classic windings screen printed windings have a much smaller geometry to reach similar ampere turns. The reason for this is the high current density of up to 100 A/mm^2 the windings can be used with. The reason which makes this possible is the proportion of cross-sectional area to surface that is responsible for a much better heat dissipation.

Another advantage of screen printed windings is the reduction of construction volume for the end winding. It is possible to reduce its volume nearly completely. In Fig. 5 the step of assembling a screen printed winding to a conventional stator back iron can be seen. It is usable with linear and rotary machines as well. After assembling, the end windings can be turned down. Therewith nearly no space is necessary for the end windings in axial machine direction.

Figure 6 is showing the construction volume of the end windings for different winding technologies. Especially compared with distributed windings much space for the end windings can be economized. Another benefit is the better utilization of space in the slot without the use of special pre-formed windings.

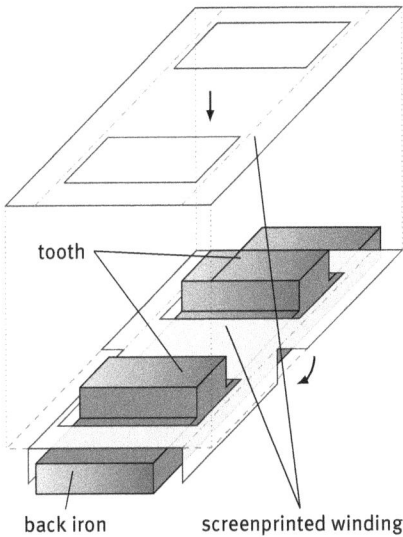

tooth

back iron   screenprinted winding

Fig. 5. Assembling process of a screen printed winding to a conventional stator back iron. It can be used for linear and rotary machines.

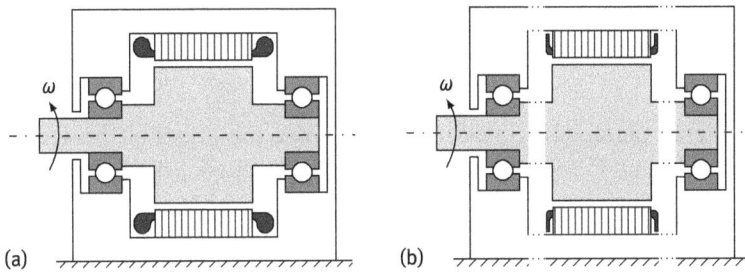

(a)   (b)

Fig. 6. Reduction of necessary space for the winding head by using screen printed windings (b) in comparison to classic windings (a) without pre-formed windings.

Beyond the mentioned benefit the screen printed windings are mechanically as robust as known from conventional copper windings. Once installed and fixed with glue the winding is not able to move caused by electromotive force in any direction.

In this paragraph a calculation should show the advantages of screen printed windings in comparison to classic distinct phase windings and tooth coils. In order to have an objective comparison the number of windings w_s and the nominal current I_N are kept constant in all variants. Starting point for the calculations should be the power dissipation and thermal behaviour of the screen printed winding. Table 2 shows all known parameters. The resistivity of the screen printed winding is an experimental value and depends on the screen printing process as well as the precise geometry of the printed conductors.

Tab. 2. Characteristic Winding Parameters.

parameter	Symbol	value
specific electric resistivity	ρ_{Ag}	$1,587 \cdot 10^{-2}\ \Omega mm^2/m$
specific electric resistivity	ρ_{Cu}	$1,678 \cdot 10^{-2}\ \Omega mm^2/m$
specific electric resistivity	$\rho_{Ag(sp)}$	$72,6 \cdot 10^{-2}\ \Omega mm^2/m$
rated temperature rise	$\vartheta_{End\text{-}Cu}$	155 K
rated temperature rise	$\vartheta_{End\text{-}Ag(sp)}$	300 K
thickness of silver paste	$h_{Ag\text{-}sp)}$	0.015 mm
thickness of galvanic copper	$h_{Cu\text{-}sp)}$	0.010 mm
thickness of dielectric	$h_{dielectric\text{-}sp}$	0.010 mm
widths of conductor	w_{sp}	2.000 mm
distance between conductors	a_{sp}	0.200 mm
space factor of copper	φ_{Cu}	0.6

Because of the higher permissible rated temperature rise the losses of the screen printed winding can be up to 1.94 times higher than this of classic copper windings under condition of an equal surface and specific heat dissipation. To handle this high temperatures a ceramic foil is used as a substrate. The surface responsible for heat dissipation is nearly the same with screen printed windings in comparison to classic copper windings. Beyond this, screen printed windings have an up to 10 times better specific heat dissipation because of their fixing with heat transfer paste in the slot as well as the end windings folding of the back iron. Therewith, good heat dissipation and a reduction of construction volume of the winding head in comparison to classic windings especially distinct phase windings are possible.

With help of Fig. 7, equation 5 gives the cross-sectional area of a screen printed conductor in comparison to a classic copper wire. After the printing process the structures were electroplated with copper. Therewith it is possible to reduce the resistance appreciable.

$$\frac{A_{sp}}{A_{Cu}} = \frac{\left(h_{Ag(sp)} + h_{Cu(sp)}\right) \cdot w_{sp}}{0.25\ mm^2} = 0.2 \tag{5}$$

The cross-sectional area of the copper depends on the diameter of the conductor that is varied between 0.25 mm² and 0.75 mm² [12]. For all following calculations the

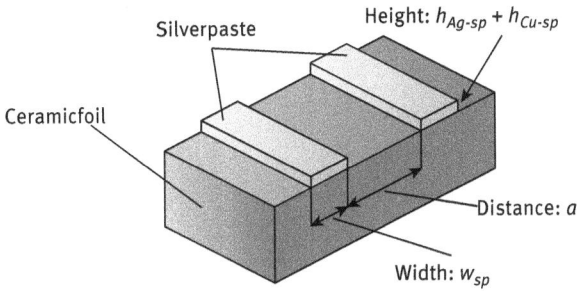

Fig. 7. Schematic of the dimensions of the screen printed windings.

worst case was used. Therewith, the relation between the resistance of screen printed windings and classic copper wire windings can be calculated with equal conductor length. Equation 6 shows that the resistance of the screen printed windings is about 12 times higher than this of classic windings.

$$\frac{R_{sp}}{R_{Cu}} = \frac{\left(\dfrac{1}{\rho_{Ag\text{-}sp} \cdot \dfrac{l_{sp}}{A_{Ag\text{-}sp}}} + \dfrac{1}{\rho_{Cu\text{-}sp} \cdot \dfrac{l_{sp}}{A_{Cu\text{-}sp}}}\right)^{-1}}{\rho_{Cu} \cdot \dfrac{l_{Cu}}{A_{Cu}}} = 12.08 \tag{6}$$

Against a background of the same phase current the relation of the ohmic losses is the same than this of the resistance.

Now it is possible to calculate the necessary construction volume of the winding in the slot and furthermore the change in machine out diameter 7. Therefore the same geometric relations as in section 2.1 were used and are given in Tab. 3.

Tab. 3. Variation of Reference Machine Parameters.

relation	value
h_y/d_y	0.03 ... 0.075
h_y/h_s	0.5 ... 1.5

$$\frac{D_{sp}}{D_{Cu}} = 1 + 2 \cdot \left(\frac{h_{s\text{-}sp}}{h_{s\text{-}Cu}} - 1\right) \cdot \frac{h_y}{h_s} \cdot \left(1 + \frac{D_r}{h_s}\right)^{-1} \tag{7}$$

with:

$$\frac{h_{s\text{-}sp}}{h_{s\text{-}Cu}} = \frac{A_{sp}}{A_{Cu}}$$

Under the condition of a constant inner diameter D_i, induction in the air-gap B_g and nominal current I_s, the torque T would also be constant. Therewith the torque density is calculated like in equation 8.

$$C \sim \frac{1}{\left(\dfrac{D_{sp}}{D_{Cu}}\right)^2} \qquad (8)$$

Design parameters for the calculation of other screen printed windings with special geometries for small-power electrical machines with air-gap windings and magnetic bearings can be found in [18] and [19].

4 Results

The increase in utilization factor of machines adopting the three examined materials FeCo, FeNi and SMC differ highly depending on the mentioned quantities and ratios. Extensive variation calculations were performed to find the maximum realistic values:

$$\text{FeCo: } C_{max} = \mathbf{165\%} \cdot C_{FeSi}$$
$$\text{FeNi: } C_{max} = \mathbf{110\%} \cdot C_{FeSi}$$
$$\text{SMC: } C_{max} = \mathbf{97\%} \cdot C_{FeSi}.$$

As stated, FeCo enables to utilize machines much higher then FeSi does. FeNi can also be lucrative. Only SMC does not allow any higher torque densities in the considered parameter ranges. This might be different at frequencies well above 1 kHz or with other machine types. However, all three materials can also lead to significantly decreased torque densities or even infeasible designs with inappropriate parameters. FeCo is only little prone to this effect but SMC highly suffers from too high temperatures because of high core loss at low frequencies. Tab. 4 outlines parameter tendencies within the ranges given in Tab. 1 of a reference FeSi machine. They describe the requirements to facilitate highest torque density improvements when utilizing different core materials.

Another method of increasing the torque density is the use of screen printed windings. According to the calculations in this article a reduction of volume of the electrical machine between 10 and 40 % is possible. The correct value is depending

Tab. 4. Necessary parameter tendencies of a FeSi machine to achieve a maximum utilization factor with different core materials.

	f	low	low	high
	H_t, H_y	low	low	high
Parameter	$w_{t,\text{FeSi}}/w_{s,\text{FeSi}}$	high	high	low
tendencies of	$h_{y,\text{FeSi}}/h_{s,\text{FeSi}}$	high	high	low
a FeSi	$h_{s,\text{FeSi}}/w_{s,\text{FeSi}}$	low	high	—
machine	$d_{y,\text{FeSi}}/h_{y,\text{FeSi}}$	—	—	—
	$P_{ct,\text{FeSi}}/P_{l2R,\text{FeSi}}$	low	high	high
	$P_{cy,\text{FeSi}}/P_{l2R,\text{FeSi}}$	high	high	—
Maximum utilization factor when FeSi exchanged with		*FeCo*	*FeNi*	*SMC*

on the exact machine geometry as well as the layout of the screen printed windings. The resulting increase of torque density is

$$C_{\text{max-sp}} = \mathbf{110...160\%} \cdot C_{\text{Cu}}.$$

At least it could be declared that the use of screen printed windings has its maximum benefit compared to machines with comparatively long teeth. Because of the higher possible working temperature, the flat structure as well as a better heat dissipation the screen printed windings are thermal stable. Beyond this it has to be kept in mind that the efficiency of screen printed windings is worse to this of classic ones. But there are many interesting approaches for a solution to get equal resistances like in conventional windings.

Combining both technologies can lead to significant higher utilization factors. Both derivations change different parameters. Varying the core material calculates new cross-areas of the soft magnetic sections leading to bigger slots and higher current with constant machine volume. Utilizing screen printed windings changes the slot height with constant current and soft magnetic cross-areas resulting in smaller machine volume. Thus, both calculations can be superimposed resulting in utilization factors of up to

$$C_{\text{max-sp,FeCo}} = 160\% \cdot 165\% \cdot C_{\text{Cu,FeSi}} = \mathbf{260\%} \cdot C_{\text{Cu,FeSi}}.$$

An example of the geometry optimization by utilizing the mentioned technologies is given in Fig. 8.

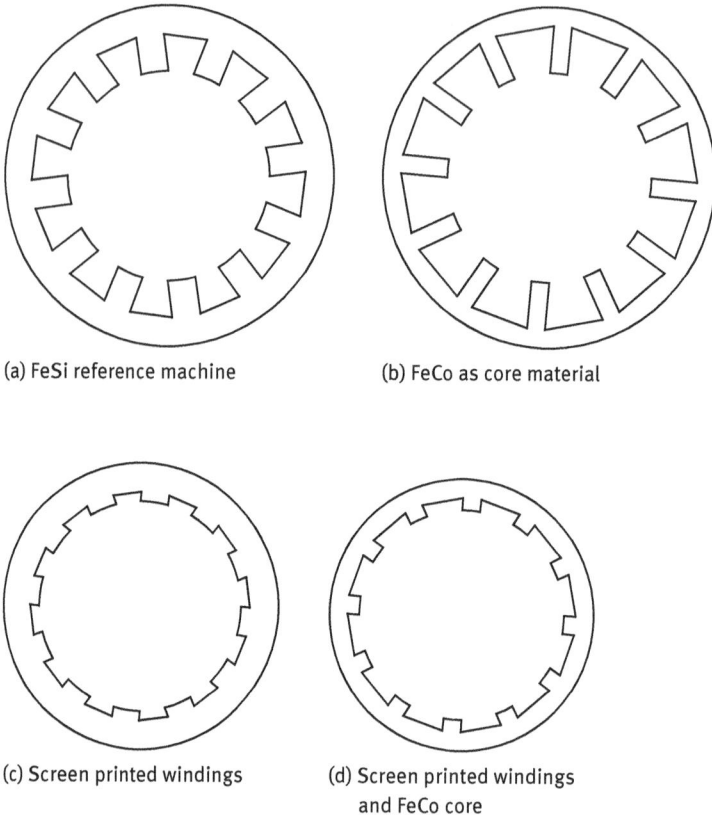

(a) FeSi reference machine

(b) FeCo as core material

(c) Screen printed windings

(d) Screen printed windings
and FeCo core

Fig. 8. Reduction of construction volume by using screen printed windings instead of classic windings and increase of slot dimensions by substituting FeSi with FeCo.

5 Conclusion

This article shows that it is possible to increase the torque density and therewith the utilization factor of electrical machines in a wide range. Therefore the influence of new materials like **FeCo**, **FeNi** and **SMC** for the construction of the core is discussed. The biggest benefit in utilization factor of up to **160** % could be achieved with the use of **FeCo** due to the higher possible inductions in the core sections. Therewith, more construction volume for the winding is possible and the torque can be increased based on a higher magnetomotive force.

Beyond this, the influence of screen printed windings on the torque density is discussed. The biggest advantage of this technology is the small construction volume and the good thermal behaviour. Thus, the overall volume of the machine can be reduced increasing the utilization factor up to **260**%. This is possible because the screen printed windings use a ceramic foil as a substrate and therewith allows a factor 1.94 higher possible winding temperature. Combined with a 10 times better heat dissipation the higher loss caused by the higher electrical resistance and higher current density is controllable.

Combining both approaches can lead to utilization factors that are hardly realizable with traditional surface-cooled machines. This leads to new freedoms of design, smaller machines, higher torques and better cooling compared to conventional technologies.

Acknowledgment: The authors would like to thank M.A. Maxi Bellmann from pmTUC. Only with her help we were able to use the screen printing technology for producing electrical windings.

This research is funded within the European Social Fund (ESF) by the Free State of Saxony and the European Union.

Bibliography

[1]   M. Lindner, R. Werner, *Optimale Antriebsdimensionierung im mechatronischen Gesamtsystem Elektromotor-Getriebe-Last (in German)*, VDI-Berichte 2138, Antriebssysteme 2011 - Elektrik, Mechanik, Hydraulik in der Anwendung, 2011.
[2]   L.G. Cravero, *Entwurf, Auslegung und Betriebsverhalten von dauermagneterregten bürstenlosen Motoren kleiner Leistung (in German)*, Dissertation, Ilmenau University of Technology, 2005.
[3]   M. Beier, *Der Ausnutzungsfaktor moderner elektrischer Maschinen in Abhängigkeit charakteristischer Maschinenparameter (in German)*, Chemnitz University of Technology, 2011.
[4]   R. Tzscheutschler, H. Olbrich, W. Jordan *Technologie des Elektromaschinenbaus (in German)*, Berlin, Germany: Verlag Technik, 1990.
[5]   L.O. Hultman, A.G. Jack, *Soft magnetic composites - materials and applications*, Proceedings on Electric Machines and Drives Conference IEMDC'03, 2003.
[6]   M. Lindner, *Untersuchung von modernen Magnetkreismaterialien und Wicklungstechnologien für energetisch hocheffiziente Antriebsmotoren (in German)*, Chemnitz University of Technology, 2009.
[7]   VAC Vacuumschmelze GmbH & Co KG, *Internal test certificates: Trafoperm N4, Vacoflux 50, Permenorm 5000V5*, 2009.
[8]   Höganäs AB, *SMC Datasheet Somaloy 700 3P*, 2009.
[9]   G. Müller, K. Vogt and B. Ponick, *Berechnung elektrischer Maschinen (in German)*, 6. Auflage Weinheim, Germany: WILEY-VCH Verlag GmbH & Co KGaA, 2008.
[10] W. Schuisky, *Berechnung elektrischer Maschinen (in German)*, Wien, Austria: Springer Verlag, 1960.
[11] G. Gotter, *Erwärmung und Kühlung elektrischer Maschinen (in German)*, Berlin, Germany: Springer Verlag, 1954.
[12] DIN (Hrsg.), *IEC 60317-0-1 - Specifications for particular types of winding wires - Part 0-1: General requirements - Enamelled round copper wire*, Beuth Verlag GmbH, 2009.

[13] D. Hanselman, *Brushless Permanent Magnet Motor Design*, 2. Edition Cranston, Rhode Island, United States of America: The Writers' Collective, 2003.

[14] H.-J. Hanke (Hrsg.), *Baugruppentechnologie der Elektronik (in German)*, Berlin, Germany: Verlag Technik, 1994.

[15] M. Fairley (Hrsg.), *RFID Smart Labels - A "How to" guide to manufacturing and performance for the label converter*, Düsseldorf, Germany: Tarsus Publishing Ltd., 2005.

[16] P. Bräuer, Th. Schuhmann, R. Werner, K. Weigelt, M. Hambsch *Wicklungen für elektrische Energiewandler (in German)*, Deutsches Patent- und Markenamt, Deutsche Patentanmeldung, 2010.

[17] P. Bräuer, T. Schuhmann and R. Werner, *Screen Printed Windings for Small-Power Electrical Machines*, The 8th France-Japan and 6th Europe-Asia Congress on Mechatronics, Yokohama 2010.

[18] P. Bräuer, R. Werner, *Herstellung von Wicklungen für rotierende Kleinantriebe mittels Siebdruckverfahren (in German)*, VDI-Berichte 2138, Antriebssysteme 2011 - Elektrik, Mechanik, Hydraulik in der Anwendung, 2011.

[19] P. Bräuer, M. Bartscht, R. Werner, *Reduzierung des Bauraums bei einem aktiven dreipoligen Radialmagnetlager unter Verwendung von siebgedruckten Wicklungen (in German)*, Proceedings on 8. Workshop Magnetlagertechnik Zittau-Chemnitz, 2011.

Biographies

Mathias Lindner has studied electrical engineering concentrating on electrical energy at Chemnitz University of Technology and received his Diploma with distinction in 2009. After working as Project Engineer at Devotek AS in Kongsberg/Norway he returned to Chemnitz to take up an employment as research assistant at the Chair of Electrical Energy Conversion Systems and Drives of Prof. Dr.-Ing. Ralf Werner at Chemnitz University of Technology. He has specialized in design and analysis of electrical machines, properties of conventional and innovative materials as well as system optimization.

Patrick Bräuer has studied mechatronics and micro technologies at Chemnitz University of Technology. He wrote his thesis with distinction and received the Diplom in 2008. Since then he has been a research assistant at the Chair of Electrical Energy Conversion Systems and Drives of Prof. Dr.-Ing. Ralf Werner at Chemnitz University of Technology. He has specialized in the design and control of electrical drives, new technologies and materials, especially the usage of 2D/3D screen printing as novel production technology.

Ralf Werner has studied electrical engineering at the University of Technology in Karl-Marx-Stadt and received his doctor's degree in 1989. Afterwards, he was employed in several positions at Chemnitz University of Technology and at EAAT GmbH, specializing in magnetic bearings and electric drives to the point of physical limits. In 2005 he was appointed to the professorship of Electric Drives at the University of Applied Sciences Mittweida. Finally, since 2009 he has headed the Chair of Electrical Energy Conversion Systems and Drives at Chemnitz University of Technology.

www.ingramcontent.com/pod-product-compliance
Lightning Source LLC
Chambersburg PA
CBHW081110220326
41598CB00038B/7293

* 9 7 8 3 1 1 0 4 4 6 1 5 9 *